PHILLIP E. HIGHSMITH

ANDREW S. HOWARD

Converse College, Spartanburg, S.C.

 SAUNDERS GOLDEN SERIES

ADVENTURES
IN PHYSICS

1972   W. B. SAUNDERS COMPANY   PHILADELPHIA   LONDON   TORONTO

W. B. Saunders Company: West Washington Square
Philadelphia, Pa.   19105

12 Dyott Street
London, WC1A   1DB

1835 Yonge Street
Toronto 7, Ontario

Adventures in Physics   ISBN 0-7216-4660-3

Print No.:   9   8   7   6   5   4   3   2   1

PREFACE

Someone once commented that the first course in college physics is taught at three levels: physics with calculus, physics without calculus, and physics without physics. The implication is that the students for whom the third type of course is designed do not have sufficient mathematical skills to understand physics. The authors disagree with this contention, and this text is an attempt to present *physics with physics* for these students. In order to bridge the void of knowledge in mathematics, graphical analysis is explained in detail at the beginning and then is used in both explanations and problem solving throughout the text. The student develops considerable skill in the graphical approach in analyzing physical phenomena.

The book is designed for a one-term course for students whose main interests lie outside the field of physics. However, we still contend that the study of physics can be an exciting adventure for any educated person in today's scientifically oriented society. We do not claim to "cover" physics in this book. The guiding principle in selecting topics for inclusion is that they should present as many of the basic concepts as possible that are relevant to an informed individual in the last quarter of the twentieth century. We have attempted to involve the student not only in the body of knowledge that is physics but also in the method that is physics. A laboratory manual of simple experiments and demonstrations has been designed to complement the text. Although most students will take this as their only course in physics, the authors would be delighted if some of them find physics such an exciting adventure that they enroll in additional physics courses.

ACKNOWLEDGEMENT

The authors wish to thank Kenneth Cheney, Pasadena City College, for his assistance and encouragement, and we particularly wish to thank Kenneth F. Kinsey, State University of New York at Geneseo, for his detailed, critical evaluation. We also thank Dr. Jerry Cromer and other colleagues of Converse College for reviewing and giving many helpful criticisms of the manuscript. Additionally, we are grateful to Miss Kay Vipperman for the preparation of the manuscript, to Mrs. Melissa Jolly for reading and giving suggestions for the final draft, and to Mrs. Pamela Kennedy for helping with the illustrations. Last, but not least, we acknowledge our debt to all our students who have given us ideas in development of the text.

PHILLIP E. HIGHSMITH

ANDREW S. HOWARD

CONTENTS

Matter in Motion

Electricity and Magnetism

Inside the Atom

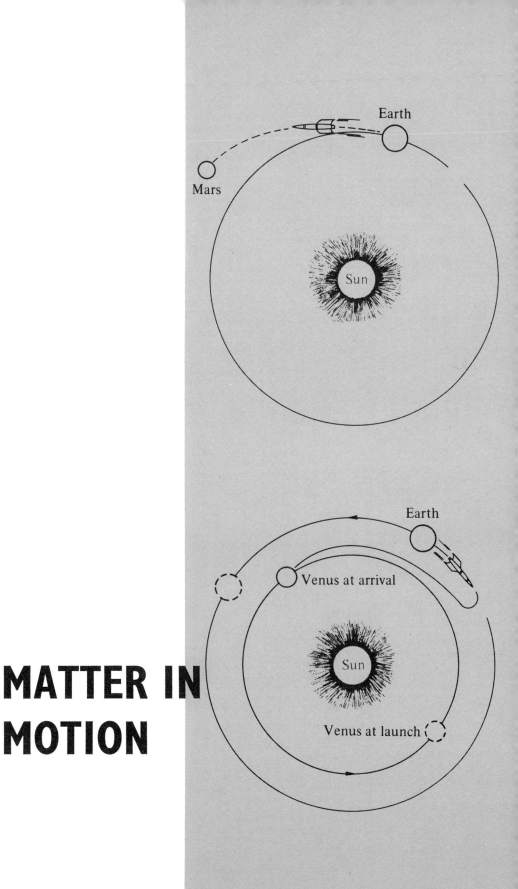

MATTER IN
MOTION

SOME FUNDAMENTAL NOTIONS

Welcome to the wonderful world of physics. It is a world that many people are afraid to look at because of rumors that it is a difficult subject, studied only by geniuses who are a little odd. Relax—one does not have to be a Mozart, a Beethoven, or a Bach to appreciate music, and neither does one have to be a Newton or an Einstein to appreciate physics. If you are willing to tackle some fundamental ideas in physics, although you may not become a physicist, you will certainly be able to understand how a physicist looks at the world.

But what is physics? A brief look at different textbooks will convince you that there are many parallel definitions of the subject. One high school student defined it as "what physicists do late at night." One college text defined it as "a science whose objective is to study the components of matter and their mutual interactions . . . ."* All the definitions are more or less correct, but incomplete. Just as the artist, the musician, or the playwright seeks to interpret the world around him in a particular way, so the physicist interprets the world from his point of view. Since the point of view is different, the results are usually different, but the truly educated person should be able to appreciate the genius of Newton, of Maxwell, and of Einstein as they interpret the world, just as one appreciates the way in which Michelangelo, Shakespeare, or Beethoven interprets the world.

The physics "thing" is to measure things in nature and, from these measurements, to predict how nature will behave. A measurement must communicate information, and this information must be comprehensible to any scientist in the field of knowledge. For example, if a person in the

---

* Alonso, M., and Finn, E. J.: Fundamental University Physics, vol. 1. Addison-Wesley Publishing Co., Reading, Mass., 1967.

United States is told that a girl's measurements are 36-24-36, he knows that his informant means inches rather than feet, yards, or miles, and he knows the areas to which the measurements apply. However, these same numbers mean little to a person who is not familiar with both the English language and girl watching.

In physics, we guarantee the clarity of measurement by defining quantities in such a way that the procedure to measure the quantity is clear to other physicists. In fact, *the definition of a physical quantity is nothing more than a description of how to measure the quantity.* Therefore, the information conveyed with a measurement tells not only how big a measurement is, but also what it is.

Measurements of physical quantities must convey at least two items of information, and can convey more. The items that must be conveyed are (1) the magnitude, or the size of the measurement (numbers are used to convey this information) and (2) the unit in which we are measuring. . For example, in the measurement of 100 seconds, "100" denotes the size of the measurement, and "seconds" is the unit of the measurement.

To describe a physical measurement, we must use some system of units. To describe a person's height as 6 is incomplete and the meaning is unclear, but to describe a person's height as 6 feet gives a complete description. Therefore, a complete description must have at least a magnitude (size) and a unit (what we are sizing).

We have a great deal of freedom in choosing units. It is equally correct to say the height of a tree is 32 feet, or 980 centimeters, or 9.8 meters, although you can probably visualize 32 feet much better than you can 980 centimeters or 9.8 meters. Unfortunately, familiarity is the only advantage of the English system of units.

FUNDAMENTAL QUANTITIES AND FUNDAMENTAL UNITS

Physicists must (and do) agree upon some system of measuring things. A very few quantities are considered fundamental and all other quantities are expressed in terms of these fundamental quantities. For each fundamental quantity, a fundamental standard unit is defined, and great pains are taken to provide accurate, available, and invariable standard units. There are three systems in which each fundamental quantity has a fundamental unit, namely, the meter-kilogram-second (mks), the gram-centimeter-second (cgs), and the slug-foot-sec (English) systems. In this book we shall use only the following fundamental quantities, with the meter-kilogram-second system of standard fundamental units.

TABLE 1–1 MKS SYSTEM

FUNDAMENTAL QUANTITY	FUNDAMENTAL UNIT
Length	Meter
Mass	Kilogram
Time	Second
Electric charge	Coulomb or ampere second

Most big basketball players are 2 meters high (1 meter ≈ 39.37 inches) and most football players have a mass of 100 kilograms (1 kilogram weighs approximately 2.2 pounds). We will define the coulomb in detail in the study of electricity.

Everything that is measured in physics is defined in terms of a fundamental quantity or a combination of fundamental quantities. If there is more than one fundamental quantity connected with the measurement, the measurement is called a derived quantity. All of these are derived quantities:

$$\text{speed} = \frac{\text{length}}{\text{time}}; \quad \text{area} = (\text{length})(\text{length});$$

$$\text{work} = \frac{(\text{mass})(\text{length})(\text{length})}{(\text{time})(\text{time})}.$$

In these examples, we say that speed has the *dimensions* of length/time, area has the *dimensions* of length^2, and work has the *dimensions* of (mass)[(length)^2/(time)^2].

The dimensions actually tell us what a measurement is or what we are measuring. For example, a snail might crawl at a speed of 1 inch/ 1 hour or 1 inch/1 day or 1 foot/minute or 2 centimeters/second.* The common thing about all these measurements is that the speed is expressed with the dimensions of length/time, and no other quantity can have these dimensions (if the dimensions are length/time, we are talking about speed and nothing else). The entire study of physics is concerned with the many different combinations of dimensions that help us interpret and measure the physical aspects of our universe.

Many combinations are given special or popular names that are really the language of physics, and knowing the language, we can reduce any problem to some combination of the fundamental quantities.

Example 1:   Do you think 5 miles north/hour is speed, or some other quantity?

Answer: Some other quantity, since it has an extra dimension.

Example 2:   How many possible quantities do you think we could have, using any or all of the four fundamental quantities in any possible combination?

Answer: Using dimensions to any power, the combinations would be unlimited.

---

* Actually, to be dimensionally correct, all units should be written in the singular. For example, 10 meter, 50 second.

GRAPHICAL ANALYSIS

It would be a wonderful world if Mother Nature were completely simple, but (like most interesting women) she is not. It is her complexities, however, that make this beautiful lady so very alluring, and the study of her secrets is a never-ending and time-consuming task, but a thrilling one. A complete understanding of the concept of graphs, which we present in the next few pages, will provide the background you need to study and scrutinize this fabulous lady in some detail. Unless you have an abundance of mathematics at your disposal, graphical analysis is the only way you can understand non-trivial physics.

A graph is a pictorial relationship between quantities that are related in some fashion. The experienced eye can tell at a glance how quantities on a graph are related. In fact, for most simply related quantities, anyone can do this. For example, Graph 1–1, (Figure 1–1*) shows a very simple relationship. It could represent the distance you travel from the center of your bed while you are sleeping. We say we have plotted distance versus time or the distance as a function of time in Graph 1–1. As the time increases, you remain in your bed.

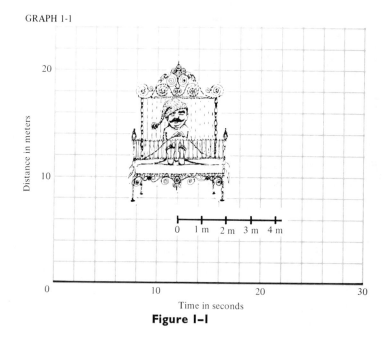

Figure I–I

---

* Every graph has been given a figure number as well as its graph number. This enables us to number the figures consecutively throughout each chapter.

Example 1:   How far are you from your bed (a) at 20 seconds? (b) at 30 seconds?

              Answer: (a) 0 meters. (b) 0 meters.

Example 2:   Looking at Figure 1–1, how far would you predict you would be from your bed in (a) 60 seconds? (b) 3600 seconds?

              Answer: (a) Most people take more than a minute to get up, so a prediction of 0 meters is okay. (b) Prediction is anyone's guess; we need more data about the subject's sleeping habits.

The next order of complication is the graph in Figure 1–2, which describes an object that is at a constant distance from a starting point (for the first 600 seconds). Notice that as the time increases, the distance does

GRAPH 1-2

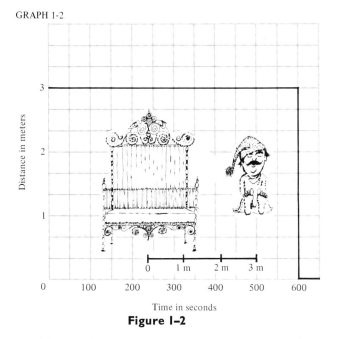

Figure 1–2

not change. This could be the graph of a person who prefers to sleep on the floor, 3 meters from the center of his bed. In graphs like the one in Figure 1–2, time is measured along the horizontal axis, called *the abscissa,* and distance is measured along the vertical axis, called *the ordinate.*

Example: How far are you from your bed (a) at zero seconds? (b) at 400 seconds? (c) What do you think took place at 600 seconds?

Answer: (a) 3 meters. (b) 3 meters. (c) The fastest trip back to bed in recorded history.

The next order of complication is shown in Figures 1–3, 1–4, and 1–5, each of which describes an object that passes a mark called distance = 0 when a timer is started at time = 0.

For example, imagine you are on a bridge over a three-lane interstate highway and three cars are abreast as they pass under you. You start your stopwatch, and for 5 seconds you plot the distance as a function of time. You record the results for car 3 on the graph in Figure 1–3, for car 4 in Figure 1–4, and for car 5 in Figure 1–5. Notice that all three graphs plot distance as a function of time, all three start at an origin, and all three plot a straight line.

There are also some things that are different about the three graphs. Graph 1–3 describes a car that travels 10 meters in 5 seconds, Graph 1–4 describes a car that travels 25 meters in 5 seconds, and Graph 1–5 describes a car that travels 50 meters in 5 seconds. Now, a little observation will convince you that Graph 1–5 is steeper than Graph 1–4, which in turn is steeper than Graph 1–3. The steepness of the curve* is related to

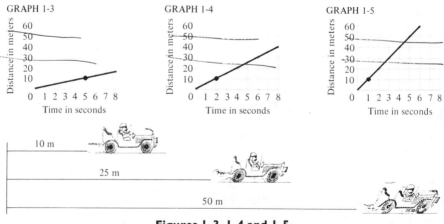

Figures 1–3, 1–4 and 1–5

* A curve is any set of connected points; therefore, a curve can be a straight line or any other shape.

how fast the distance is changing each second. Therefore, we need a measure for the steepness or *slope* of a curve. The following criteria would seem desirable if we were to devise a "slopemeter" to measure the slope:

(1) If the curve goes uphill, the slope should be positive and the slopemeter should give a positive reading.
(2) If the curve goes downhill, the slope should be negative and the slopemeter should give a negative reading.
(3) If the curve is level, the slope should be zero and the slopemeter should read zero.
(4) If the curve is steep, the value of the slope should be large; in fact, the magnitude (size) of the slope should be a direct measure of how steep the curve is.

GRAPH 1-6 illustrates the criteria

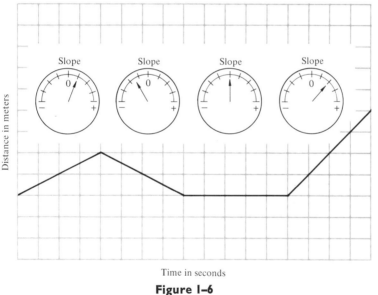

Time in seconds

Figure 1–6

The next order of complication is a graph like that in Figure 1–7, which describes an object that starts from rest and continuously gains speed. In this case, the object could be a car in a drag race. Note that the slope of the curve as well as the distance changes at each instant.

Mathematical curves are very important in physics, since so many "laws of nature" are too complex for simple mathematical description, and are described by graphs similar to the one in Figure 1–7. You will study in some detail the *slope* of a curve. *Reading a graph is an important skill that you must have.*

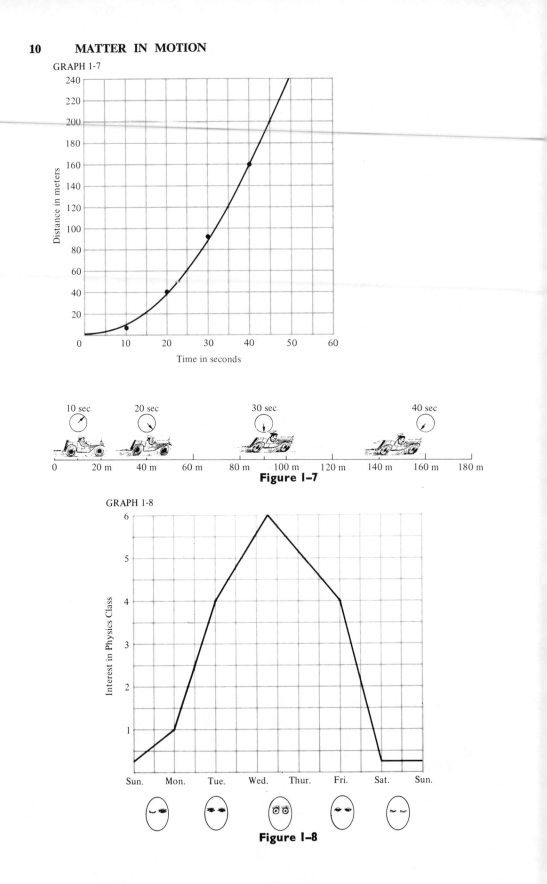

GRAPH 1-7

Figure I-7

GRAPH 1-8

Figure I-8

THE SLOPE OF THE CURVE

Figure 1–8 describes interest in physics according to the day of the week. Answer the following question, using the concepts we have already discussed:

Example 1:   What is the steepest part of the hill going up? What does this tell you?

   Answer: Monday to Tuesday. Interest in physics increases most rapidly (test on Wednesday).

Example 2:   What is the steepest part of the hill going down? What does this tell you?

   Answer: Friday to Saturday. Student interest in physics drops near the weekend, for some strange reason.

Example 3:   What part of the curve is flat or has zero slope?

   Answer: Saturday to Sunday.

Example 4:   If there is day in the week when there is no interest in physics?

   Answer: No, the thoughts of homework due on Monday prevent this.

We need a more precise measuring device to measure the slope of a curve at any point. The slope of a curve is defined in the following manner:

$$\text{Slope} = \frac{\Delta \text{ Quantity plotted vertically}}{\Delta \text{ Quantity plotted horizontally}} = \frac{y_2 - y_1}{x_2 - x_1} = \frac{\Delta y}{\Delta x},$$

where y represents any quantity plotted on the vertical axis and x is the quantity plotted on the horizontal axis. The symbol Δ (delta) means "change in," *not* a product. That is, Δx means "change in x" and Δy means "change in y." To obtain the change in x and the change in y, we can use any two sets of points, as Figures 1–9 and 1–10 illustrate. A set of points is denoted by (x, y). That is, the value along the horizontal axis is written first, followed by the value along the vertical axis.
   Figure 1–9 demonstrates a positive slope.

GRAPH 1-9

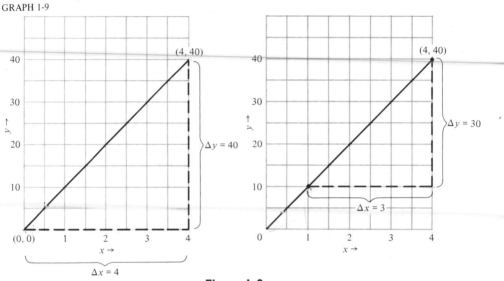

Figure 1–9

$$\begin{bmatrix} (y_2 = 40)(y_1 = 0) \\ (x_2 = 4)(x_1 = 0) \\ \Delta y = 40 \text{ units} \\ \Delta x = 4 \text{ units} \\ \text{Slope } \left(\dfrac{\Delta y}{\Delta x}\right) = \dfrac{40}{4} = 10 \end{bmatrix} \quad OR \quad \begin{bmatrix} (y_2 = 40)(y_1 = 10) \\ (x_2 = 4)(x_1 = 1) \\ \Delta y = (40 - 10) = 30 \\ \Delta x = (4 - 1) = 3 \\ \text{Slope} = \left(\dfrac{30}{3}\right) = 10 \end{bmatrix}$$

The slope is $+10$. Note that for a positive slope the curve is uphill going from left to right.

Figure 1–10 shows a negative slope.

$$\begin{bmatrix} y_2 = 0 \quad y_1 = 40 \\ x_2 = 4 \quad x_1 = 0 \\ \Delta y = y_2 - y_1 = -40 \\ \Delta x = x_2 - x_1 = +4 \\ \dfrac{\Delta y}{\Delta x} = \dfrac{-40}{4} = -10 \end{bmatrix} \quad OR \quad \begin{bmatrix} y_2 = 10 \quad y_1 = 40 \\ x_2 = 4 \quad x_1 = 1 \\ \Delta y = y_2 - y_1 = -30 \\ \Delta x = x_2 - x_1 = +3 \\ \dfrac{\Delta y}{\Delta x} = \dfrac{-30}{3} = -10 \end{bmatrix}$$

The slope is -10. Note that the curve goes downhill. If any given curve is a straight line, the slope will be constant and the value of the slope will be the same no matter what two sets of points we choose. *By defining the slope as we did, we have all the desirable criteria for measurement.* In

GRAPH 1-10

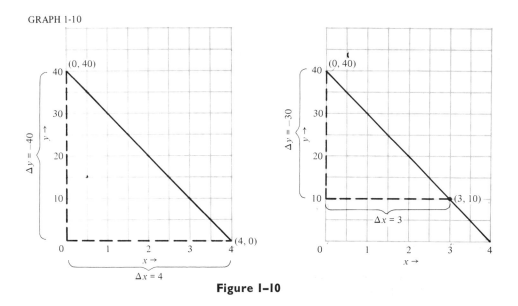

Figure I–10

traveling from left to right, if the curve goes uphill the slope is positive, and if the curve goes downhill the slope is negative. If the hill is steep, the slope has a large value, which conveys the information that a small change in the variable plotted horizontally will cause a large change in the variable plotted vertically. A positive slope tells us that when one quantity is increasing, the other is increasing. A negative slope tells us that when one quantity increases, the other quantity decreases. A straight line will have a non-changing or constant slope. If the curve is level, the slope is zero.

A curved line such as the one in Graph 1–11 (Figure 1–11) does not

GRAPH 1-11

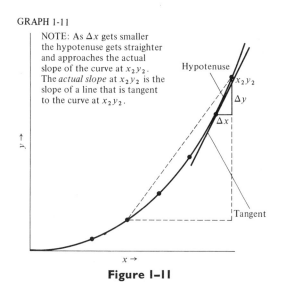

NOTE: As Δx gets smaller the hypotenuse gets straighter and approaches the actual slope of the curve at $x_2 y_2$. The *actual slope* at $x_2 y_2$ is the slope of a line that is tangent to the curve at $x_2 y_2$.

Figure I–11

have a constant slope. Therefore, we cannot give the slope except for a given point. The slope is gradually increasing, as can be seen by observing that $\Delta y/\Delta x$ is increasing. A problem arises because the hypotenuse of each "triangle" formed by $\Delta y/\Delta x$ is not a straight line, but curved. The smaller Δx is, however, the straighter the hypotenuse will be (see Graph 1–11), so we can find an excellent approximation to the slope at any point by letting Δx be very, very small. In fact, in calculus we learn that if we let Δx approach zero, the ratio of $\Delta y/\Delta x$ will be the exact slope at any given point.*

THE SPEED OF A PARTICLE

Since you have to some extent mastered the concept of the slope, you can now learn about the speed of a particle. Speed is defined as the change in distance divided by the change in time. That is,

$$\text{Speed} = \frac{\Delta \text{ distance}}{\Delta \text{ time}} = \text{slope of a distance versus time curve.}$$

Actually, there are three different speeds that have meaning: constant speed, average speed, and instantaneous speed. In order to understand these concepts, let's give examples of each type of speed.

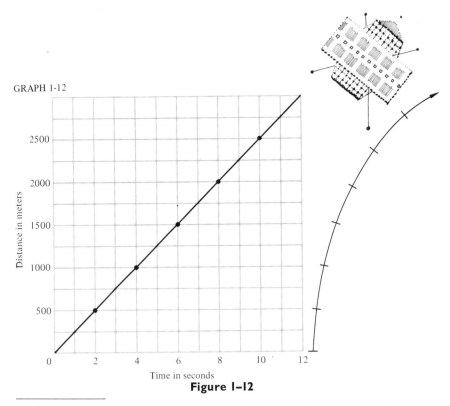

GRAPH 1-12

Distance in meters

Time in seconds

Figure I–12

---

* It is recommended that Experiment 1 in the lab manual be done at this time.

Constant Speed. Suppose that a satellite is traveling along a given path at 250 meters per second, which is approximately equal to 560 miles per hour. The distance traveled from a point along the path would increase at a constant rate, and if you were to graph the situation it would be like Graph 1–12 in Figure 1–12. Whenever the slope of a distance versus time curve is constant, the speed is constant. Notice by reading the graph the satellite travels 2500 meters in 10 seconds.

Average Speed. A rocket starts from rest and travels 2500 meters in 10 seconds. The average speed for the 10 seconds is 250 m/sec, although we know little about any part of the trip. If the distance as a function of time makes a curve (like Graph 1–13, for example), we would know much more about the situation. The slope in Graph 1–13 is not constant. The

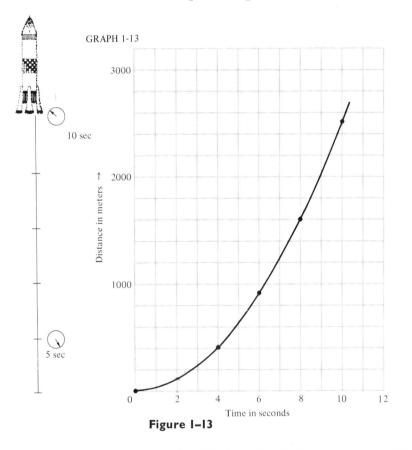

Figure I–13

average speed for any interval is still given by Δ distance/Δ time. The average speed for:

(a) the first 2 seconds $\approx \dfrac{(100 - 0)\ \text{m}}{(2 - 0)\ \text{sec}} = \dfrac{100\ \text{m}}{2\ \text{sec}} = 50\ \text{m/sec}^*$

_____

* Note that $\approx$ means approximately.

(b) the first 4 seconds $\approx \dfrac{(400 - 0) \text{ m}}{(4 - 0) \text{ sec}} = \dfrac{400 \text{ m}}{4 \text{ sec}} = 100 \text{ m/sec}$

(c) the first 6 seconds $\approx \dfrac{(900 - 0) \text{ m}}{(6 - 0) \text{ sec}} = \dfrac{900 \text{ m}}{6 \text{ sec}} = 150 \text{ m/sec}$

(d) the first 8 seconds $\approx \dfrac{(1600 - 0) \text{ m}}{(8 - 0) \text{ sec}} = \dfrac{1600 \text{ m}}{8 \text{ sec}} = 200 \text{ m/sec}$

(e) the first 10 seconds $\approx \dfrac{(2500 - 0) \text{ m}}{(10 - 0) \text{ sec}} = \dfrac{2500 \text{ m}}{10 \text{ sec}} = 250 \text{ m/sec}$

The average speed is based upon only two points of measurement. We could have taken any other two points on the graph to find the average speed during a time interval. For example, the average speed between $t = 6$ sec and $t = 8$ sec is

$$\approx \dfrac{(1600 - 900) \text{ m}}{(8 - 6) \text{ sec}} = \dfrac{700 \text{ m}}{2 \text{ sec}} = \dfrac{350 \text{ m}}{\text{sec}}$$

Instantaneous Speed.   Imagine that we wanted to know the "exact" speed of the rocket at any given time in Graph 1–13. We would want the instantaneous speed. If we let the two points on the graph get closer and closer to each other, we get a better and better approximation of the instantaneous speed at a particular time. In fact, if we let the change in time approach zero (but never actually reach zero), we have *instantaneous speed*. Instantaneous speed is defined as Δ distance/Δ time (as the change in time approaches zero).

Let's find the instantaneous speed at the end of the second, sixth, and tenth seconds. Instantaneous speed is the slope of the curve at the particular instant. Draw a line tangent to the curve at the given point and estimate the slope.*

Graph 1–14 is a replica of Graph 1–13 with the tangent lines drawn at the times we want the instantaneous speed, v. We can approximate the tangent lines for each time by letting Δt be small (for example, two seconds) and letting t_2 and t_1 be equal increments above and below each time we want the instantaneous speed.

At the end of the second second, $v_i = \text{slope} = 100$ m/sec. At the end of the sixth second, $v_i = 300$ m/sec. At the end of the tenth second, $v_i = 500$ m/sec.

You might think that if we let the change in time approach zero, the result would be zero, but this is not true. What we get is a true value of the slope of the line at a particular point or, in this case, the instantaneous speed of the particle at a particular time. The speedometer of a car is an

---

* By drawing the tangent line, we can estimate the instantaneous speed. Only by calculus can we find the *exact* instantaneous speed.

GRAPH 1-14

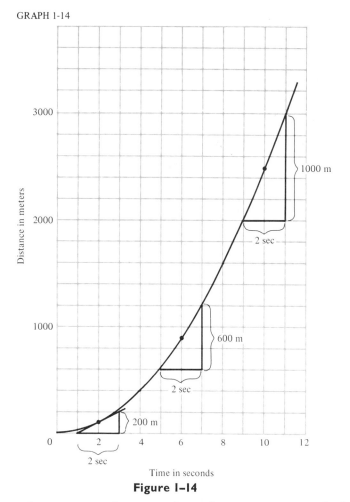

Figure I-14

instrument that measures the approximate instantaneous speed of the car; therefore, the speedometer is nothing more than a slopemeter that reads only positive values.

As long as an object travels in a straight line, speed is a sufficient concept to describe the motion. However, if an object changes *direction* (or speed), a push or pull has acted upon the object. We need a concept to incorporate this push or pull into our description of nature, so we define a term called *velocity* to include the notion of any change in the direction of the motion as well as any change in the magnitude of the motion. *In short, we define velocity to include magnitude, dimensions, and direction.* We call any quantity that includes direction a vector quantity. We call a quantity with magnitude and dimensions only (like speed) a scalar quantity. Before we study velocity, we must digress and study some of the general properties of vectors.

VECTORS

A vector quantity is always specified by a magnitude, a unit, and a direction. A vector can be represented by an arrow—the length of the arrow represents the size or magnitude of the quantity and the direction of the arrow points in the direction of the "line of action" of the vector. Figure 1–15 shows examples of vectors:

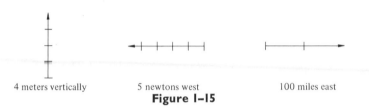

4 meters vertically   5 newtons west   100 miles east

Figure 1–15

Vectors are usually denoted by printing a V in boldface (**V**) or putting an arrow above it ($\vec{V}$).

It is convenient to give the following facts concerning vectors:

(1) Vectors are equal if they have the same magnitude and direction. (This means that we can move vectors around.)

(2) A vector multiplied by a scalar is a vector: 5(5 m north) = 25 m north.

(3) A vector divided by a scalar is a vector: 25 m north/5 sec = 5 m north/sec.

(4) A negative vector is a vector of equal magnitude as a positive vector, but in the opposite direction: (5 meters east ⟶) and (−5 meters east ⟵).

(5) A zero vector is a vector with zero magnitude and no specific direction.

(6) Two or more vectors can be added: Place the tail of one of the vectors at any convenient point. Next place the tail end of the second vector on the head of the first, and the tail of the third on the head of the second, and continue this process. The result (resultant) of this addition is the vector from the tail of the first to the head of the last. Figure 1–16 illustrates the method. It does not matter in what order the vectors are added.

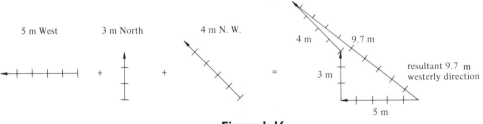

5 m West   3 m North   4 m N. W.   4 m   9.7 m

                                                         3 m   resultant 9.7 m
                                                                      westerly direction
                                                              5 m

Figure 1–16

Change in a Vector, "ΔV." Remember that Δx was defined as $(x_2 - x_1)$ and Δy as $(y_2 - y_1)$. The change in a vector ΔV is also defined as $(V_2 - V_1)$, but we must subtract the vector V_1 from the vector V_2. To subtract the vector V_1, we simply "vectorially add" the negative of the vector V_1. In short, we rotate V_1 to the opposite direction and then add the vectors together. For example, if V_1 is equal to 3 units north and V_2 is equal to 4 units west, what is the value of ΔV?

Answer: From Figure 1–17, $\Delta V = 5$ units southwest.

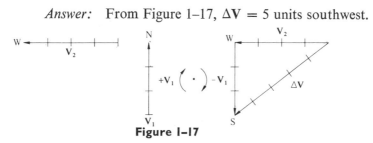

Figure 1–17

Components of a Vector. Many times it is convenient to break a vector down into two or more vectors. These vectors are called the components of the vector. Usually we break vectors into components that are perpendicular to each other and called rectangular components.

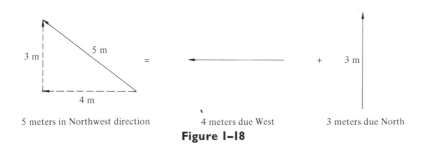

5 meters in Northwest direction   4 meters due West   3 meters due North

Figure 1–18

VELOCITY

Now you can understand the concept of velocity. The velocity of a particle is the change in position of the particle divided by the time it takes the particle to change its position. For example, suppose that a particle moves from position A to position B. If we measure all positions from point 0, the change in position from A to B is the displacement vector $0B$ (labeled D_2 in Figure 1–19) minus the displacement vector $0A$ (labeled D_1).

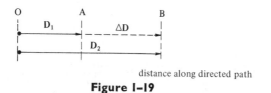

distance along directed path

Figure 1–19

We can define the average velocity of the particle as:

$$V = \frac{\Delta D}{\Delta t}$$

where ΔD is the change in displacement
and Δt is the time required to make the change.

By definition, $\Delta D = D_2 - D_1$, and if D_1 is equal to zero, ΔD is equal to the displacement D_2. In straight line motion, when we speak of the "displacement from a zero position," or more simply the "displacement," we are implying that the vector $D_1 = 0$.

Since the change in *displacement* is a vector, *velocity* is also a vector, since a vector (Δ displacement) divided by a scalar (Δ time) is by definition a vector.

The length of the vector can be drawn to scale to represent the magnitude of the displacement. Figure 1–20 illustrates displacement vectors.

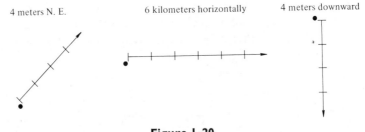

4 meters N. E.   6 kilometers horizontally   4 meters downward

Figure 1–20

A little knowledge about vectors will help you understand the difference between the speed of an object and the velocity of an object. For example, suppose that an object travels 40 kilometers east from a zero position in 8 hours and 30 kilometers north in 2 hours. What are the average speed and the average velocity for the entire trip?

Solution: $\text{Speed} = \dfrac{\Delta \text{ distance}}{\Delta \text{ time}} = \dfrac{(40 + 30) \text{ km}}{(8 + 2) \text{ hr}}$

$$= \frac{70 \text{ km}}{10 \text{ hr}} = 7 \text{ km/hr}$$

$$\text{Velocity} = \frac{\Delta \text{ displacement}}{\Delta \text{ time}} = \frac{(50 - 0) \text{ km N.E.}}{10 \text{ hr}}$$

$$= \frac{5 \text{ km N.E.}}{\text{hr}}$$

Note in Figure 1–21 that the average speed and the average velocity have different magnitudes. Only when an object is traveling in a *straight line* is the magnitude of average velocity equal to the average speed.

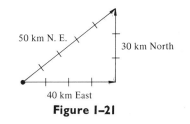

50 km N. E.   30 km North

40 km East

Figure 1–21

Instantaneous Velocity. Instantaneous velocity is defined as the ratio of the change in displacement to the time interval necessary to make the change in the displacement as the time interval approaches zero. That is,

Instantaneous velocity $V_i = \Delta$ displacement$/\Delta$ time

(as Δt approaches zero).

The magnitude of the instantaneous velocity is the instantaneous speed of an object. This is probably why so many people use the terms velocity and speed interchangeably.

The direction of the instantaneous velocity is that of a tangent to the path. For example, if a rock is going in a circle with an instantaneous speed of 30 m/sec, the instantaneous velocity is 30 m/sec. The direction at any instant can be found by drawing a tangent to the circle at the position of the rock at that instant.

The following examples will clarify these concepts.

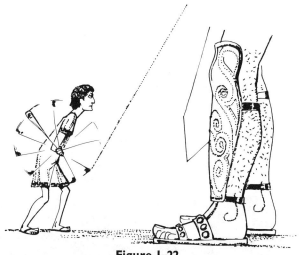

Figure 1–22

Example 1:   Figure 1–23 shows the displacement of a rocket rising vertically at take-off in a straight line. (NOTE: "Straight line" also implies a given direction.) What is the instantaneous velocity at $t = 4$ seconds?

Answer: Instantaneous velocity $(V_i) = \Delta D/\Delta t$ (as t approaches zero) = slope at $t = 4$ seconds. A good approximation is to take as small a time interval as possible. Using a one-second time interval, the slope = $V_i =$ 20 m/1 sec (upward in a straight line).

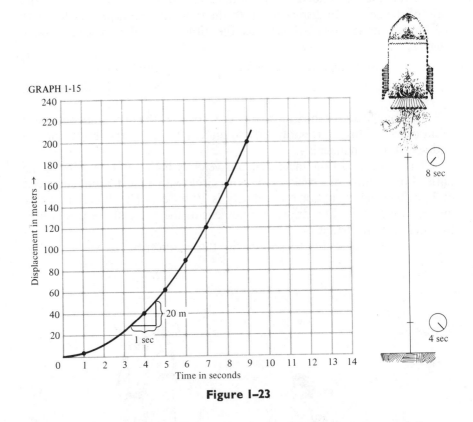

GRAPH 1-15

Displacement in meters →

Time in seconds

Figure I–23

Example 2:   What is the average velocity for the first 4 seconds?

Answer: Average velocity $V_a = \Delta D/\Delta t$. Reading from the graph, $\Delta D = 40$ m (in straight line), and $\Delta t = 4$ seconds

$$V_a = \frac{\Delta D}{t} = \frac{40 \text{ m}}{4 \text{ sec}} = \frac{10 \text{ m}}{\text{sec}} \quad \text{(upward in straight line).}$$

Example 3:   What is the instantaneous speed at $t = 4$ seconds?

> *Answer:* Instantaneous speed = magnitude of the velocity, so
> $$V_i = \frac{20 \text{ m}}{\text{sec}}.$$

Example 4:   What is the average speed for the first 4 seconds?

> *Answer:* Since the displacement is in a straight line, the magnitude of the average velocity is equal to the speed:
> $$V_a = \frac{10 \text{ m}}{\text{sec}}.$$

These problems illustrate that as long as the motion is in a straight line, speed and velocity have equal magnitudes. In problems involving straight line motion, either speed or velocity is asked for or given; however, it should be understood that if velocity is asked for or given, the direction of motion is implied.

Relative Velocity.   Have you wondered how a spacecraft traveling toward the moon with a velocity of 11,000 meters/sec (25,000 miles/hr) could dock with another spacecraft traveling 11,000 meters/sec? This feat can only be accomplished if the velocities are in the same direction and the velocity of both spacecrafts is measured relative to the same frame of reference. In this case, the velocity of one spacecraft relative to the other is equal to zero. The pilot of each spacecraft looks at the other, and both agree they are not moving relative to each other, although each craft is moving relative to the frame of reference with a large velocity. If the object to which the velocity is relative is not mentioned, it is understood to be the Earth. *All velocities are relative.*

MORE ABOUT GRAPHS

So far we have not considered curves that are to the left of and under the origin (point 0, 0). Let's interpret the physical situation in Graph 1–16 (Figure 1–24). When we start timing, we usually press a button on a timepiece and call this time $= 0$. At $t = 0$, the object is 20 meters from a position we arbitrarily call zero. The speed of the object is constant and is equal to

$$\frac{\Delta d}{\Delta t} = \frac{30 \text{ meters}}{6 \text{ seconds}} = 5 \text{ m/sec.}$$

We can also give an interpretation to both ends of the graph if we are

GRAPH 1-16

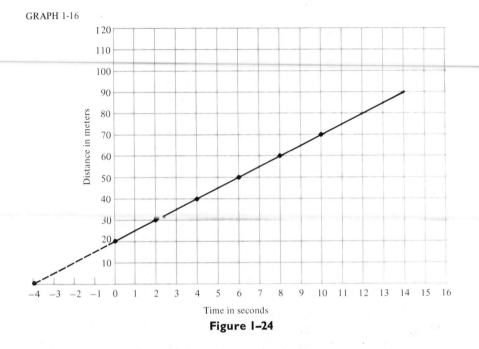

Figure 1–24

convinced that the physical situation was going on before we started timing and will continue past the time covered in Graph 1–16. If we extend the curve to the left, we find that the object passed our zero position 4 seconds before we started the timer (at $t = -4$, $d = 0$). By using negative time values, we can predict how long ago the object was at any given position. We also can predict between any data points or extend the graph beyond our data points if we are confident that the physical situation permits us to do this. Usually another experiment will justify or nullify our right to do this.

Example: (a) When was the object 10 meters from the starting position? (b) Where was the object at $t = 5$ sec and $t = 12$ sec?

Answer: (a) Extending the graph to the left, $d = +10$ m at $t = -2$ sec. Also, if we interpret $d = -10$ m as being on the opposite side of the starting position, we have $d = -10$ m at $t = -6$ sec. (b) Reading from the graph at $t = 5$ seconds, distance $= 45$ m; extending the graph to 12 sec, distance is equal to 80 m.

It is seldom that a curve will fit all data points. Therefore, we seek a smooth curve (among all possible curves) that satisfies the data. We can program a computer to provide us with the curve that best fits the data; however, in most situations the answer is quite obvious after plotting the data points. You should have about the same number of data points missing the smooth curve on each side.

If the data points indicate a straight line and you want the slope of the line, you should calculate the slope from the line and not from any two data points. In this way you get the best approximation for the slope, since you have "averaged out" your errors by drawing that curve that best fits all data points.

Looking at Graph 1–17, you can see that almost no data points fall exactly on the "best curve" but that there are almost as many points above the curve as below the curve. To find the speed, we could use the points like (0 sec, 0 m) and (11 sec, 110 m) taken from the curve rather than the exact data points (0 sec, 5 m) and (11 sec, 105 m).

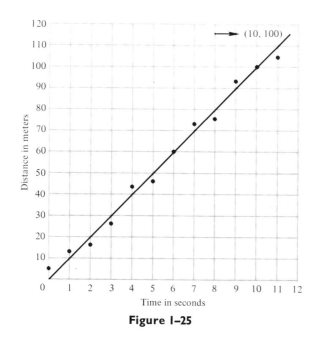

Figure I–25

LARGE AND SMALL NUMBERS

In physics we deal with very large and very small numbers—for example, the mass of the proton is 0.000000000000000000000000000167 kilograms, and the speed of light is 300,000,000 meters/sec. In order to keep mental institutions from being too crowded with scientists, the following method for handling large and small numbers is used.

(1) We recognize that our number system is based on 10 (in all probability because we have 10 fingers to count on).

(2) We define an exponent as a number, written as a superscript, that tells us how many times another number, called the base, is to be multiplied; for example,

$$10^2 = 10 \times 10 = 100 \quad (10 \text{ is the base, 2 is the exponent})$$
$$10^3 = 10 \times 10 \times 10 = 1000$$
$$10^4 = 10 \times 10 \times 10 \times 10 = 10,000$$

(3) Any base raised to the zero power is equal to 1.

Examples:
$$10^0 = 1 \qquad 3^0 = 1 \qquad (x + y)^0 = 1 \qquad A^0 = 1$$

(4) Large numbers are written as a small number times 10 to some power. It is much simpler to keep up with decimal points in this manner.

Examples:
$$5280 = 5.280 \times 10^3 = 52.80 \times 10^2 = 528.0 \times 10^1$$
$$240,000 = 2.40 \times 10^5$$
$$68,000 = 6.8 \times 10^4$$
$$529 = 5.29 \times 10^2$$
$$18,000 = 18 \times 10^3$$

(5) We define a negative exponent as the reciprocal of a positive exponent.

Examples:
$$1/100 = 1/10^2 = 10^{-2}$$
$$1/1000 = 1/10^3 = 10^{-3}$$
$$1/10,000 = 1/10^4 = 10^{-4}$$
$$1/10^6 = 10^{-6}$$

In order to work conveniently with small numbers, we use the rule that multiplication by one does not change a quantity. Therefore, if we wish to move the decimal point, we multiply the number by a power of 10 and divide by the same power of 10. This enables us to move the decimal to our liking. In scientific notation, a number is written as a decimal number between one and 10, multiplied by some power of 10. The following examples illustrate the method:

Multiplication by 1000 $\times$ Division by 1000

$$0.00685 = 6.85 \times 10^{-3} \ or \ 0.00685 \ \frac{(1000)}{(1000)} = \frac{6.85}{1000} = 6.85 \times 10^{-3}$$

Multiplication by 1,000,000 $\times$ Division by 1,000,000

$$0.00000785 = 7.85 \times 10^{-6} \ or \ 0.00000785 \ \frac{(1,000,000)}{(1,000,000)} = \frac{7.85}{1,000,000}$$

$$= 7.85 \times 10^{-6}$$

It is easier to compute numbers in scientific notation:

Example:

$$\frac{5,000,000}{250} = \frac{500 \times 10^4}{250} = 2 \times 10^4$$

$$\frac{0.00500}{250} = \frac{500 \times 10^{-5}}{250} = 2 \times 10^{-5}$$

SIGNIFICANT NUMBERS

The preceding method also gives us a way to communicate the accuracy of measurement. The numbers of places of accurately known digits or significant figures are always written as a mixed decimal number between 1 and 10, and the powers of 10 give the correct location of the decimal point.

Examples: 530,000 correct to two significant figures is 5.3×10^5
530,000 correct to three significant figures is 5.30×10^5
530,000 correct to four significant figures is 5.300×10^5

In review, exponential notation provides a method to keep up with the decimal point by using positive and negative powers of 10. Scientific notation provides a method of communicating the accuracy of our measurement by writing the number of significant figures as a decimal number between one and 10 multiplied by 10 to some power.

Examples: 93.0×10^6 has three significant figures

1.6×10^{-24} has two significant figures

9.210×10^{-20} has four significant figures

PREFIXES

Another way in which we handle large and small numbers is to use a prefix. Whenever a prefix is used with any quantity, you should multiply the quantity by the amount indicated by the prefix. Since the following prefixes are used so frequently, you should learn them.

giga = G (multiply by 1,000,000,000 or 10^9)

mega = M (multiply by 1,000,000 or 10^6)

kilo = k (multiply by 1000 or 10^3)

centi = c (multiply by 1/100 or 10^{-2})

milli = m (multiply by 1/1000 or 10^{-3})

micro = μ (multiply by 1/1,000,000 or 10^{-6})

nano = n (multiply by 1/1,000,000,000 or 10^{-9})

pico = p (multiply by 1/1,000,000,000,000 or 10^{-12})

You can easily see that
one kilogram is 1000 grams
one centimeter is 1/100 meter
one megaton is 1,000,000 tons
one milligram is 1/1000 grams

Example 1:   The national debt of the United States at one time was 335 billion dollars. Express this in megabucks and in gigabucks. (A billion is 1,000,000,000.)

Answer: 335,000 megabucks or 335 gigabucks.

Example 2:   The largest explosive device in the world is a large hydrogen bomb that is rated as powerful as 100 megatons of conventional explosives. How many tons is this?

Answer: 100,000,000 tons.

EXERCISES

1. Do you think we should agree on one system of measurement (such as the English system) and abolish all others? Which system would you suggest we keep?

Figure 1–26 shows the distance of a lunar module from the mother spacecraft as a function of the time.

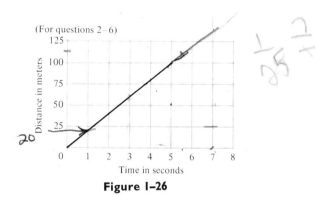

(For questions 2–6)

Figure I–26

2. How far was the module from the spacecraft at the end of $5\frac{1}{2}$ seconds?

3. Was the speed constant?   *yes*

4. What was the speed at the end of 3 seconds?

5. What was the speed at $t = 0$? $\frac{\Delta a}{\Delta t} = 0$

6. What was the distance at $t = 7$ sec?

7. What would be the value of:
   (a) 10^6?   *1.000000*
   (b) 10^1?   *10*
   (c) 10^9?   *1000000000*
   (d) 10^5?

8. What is the value of:
   (a) 1000^0?
   (b) $(10^6)^0$?
   (c) A^0?
   (d) $(X - Y)^0$?

9. (a) $68{,}000 = 6.8 \times 10^?$ *4*
   (b) $529 = 5.29 \times 10^?$ *2*
   (c) $18{,}000 = 18 \times 10^?$ *3*
   (d) $0.000527 = 5.27 \times 10^?$ *-4*

Answer the following questions about Figure 1–27.

(For questions 10–15)

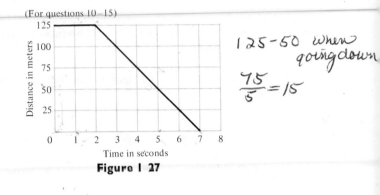

125 - 50 when going down

$\dfrac{75}{5} = 15$

Figure I 27

10. What was the distance at $t = 1$?   125 m

11. What was the speed at $t = 1$? $\dfrac{\Delta d}{\Delta t}$ $\dfrac{0}{1} = 0$ m/sec

12. How long was the object at rest?   2 sec.

13. What was the distance at $t = 5$ sec?   50 m

14. What was the speed at $t = 5$ sec?

15. What happened at 7 seconds?   stopped

16. (a) $10^{-2} = \dfrac{1}{10^{2}}$ 2   (c) $\dfrac{1}{10^{6}} = 10^{7}$ 6

    (b) $\dfrac{1}{10^{-2}} = 10^{2}$

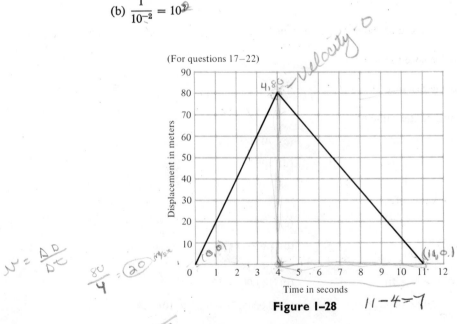

(For questions 17–22)

velocity = 0

4,80

(0,0)

(14,0.)

$v = \dfrac{\Delta D}{\Delta t}$

$\dfrac{80}{4} = 20$ m/sec

Figure I-28 $11 - 4 = 7$

$\dfrac{80}{-7}$ 11

7

Figure 1–28 shows the results of an experiment in which a ball was rolled directly toward a perpendicular wall. Answer the following questions about the motion. (NOTE: Assume straight line motion.)

17. At what time did the ball hit the wall? *4 sec.*

18. With what velocity did it approach the wall? *20 m/sec*

19. What was the rebound velocity? *–11.4.*

20. At what time or times was the ball 50 meters from the starting point? *2.5 + 6.7 sec.*

21. At what time was the velocity zero? *4 sec.*

22. How long did it take for the ball to return from the wall to the starting point? *7 sec,*

23. A boy riding a bike travels 6 kilometers due west and then 8 kilometers due north. It takes 2 hours to complete the entire trip. (a) What is his average speed and *14/2 = 7 km/hr.* (b) what is his average velocity?

24. Add the following vectors and find the resultant: 3 kilometers north, 5 kilometers east, and 4 kilometers southeast. *4+5*

Figure 1–29 shows the results of an experiment of a ball rolling straight down a hill. Answer the following questions concerning the experiment.

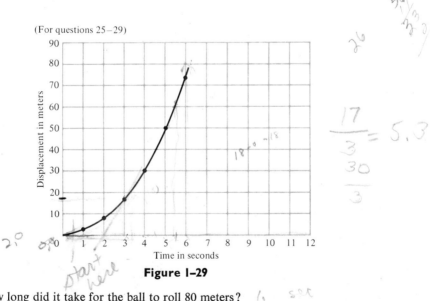

(For questions 25–29)

Figure 1–29

25. How long did it take for the ball to roll 80 meters? *6 sec*

26. Was the velocity constant? If not, was the velocity increasing or decreasing? *NO*

27. Estimate the velocity at the end of 3 seconds. *13*

28. Estimate the velocity at the end of 5 seconds.

29. What was the change in the velocity from the end of the third to the end of the fifth second?

30. What is:

(a) 5287 correct to one significant figure?

(b) 5287 correct to two significant figures?

(c) 5287 correct to four significant figures?

31. A young physicist prefers to keep a diary in graphical form. Figure 1–30 shows the entry describing his weekend picnic. Write a short paragraph telling:

(a) what the data shows,

(b) any inferences you can make from the data.

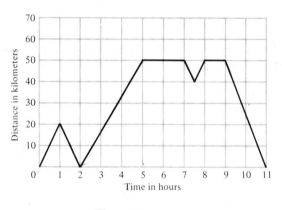

Figure I–30

Figure 1–31 shows an experiment in which an object was thrown vertically into the air. Answer the following questions concerning the experiment.

32. At what time was the velocity zero? 3 sec.

33. For the first 3 seconds, was the velocity increasing or decreasing?

34. What was the initial velocity (approximately)?

35. What was the final velocity (approximately)?

36. At what time(s) was the object 25 meters from the ground?

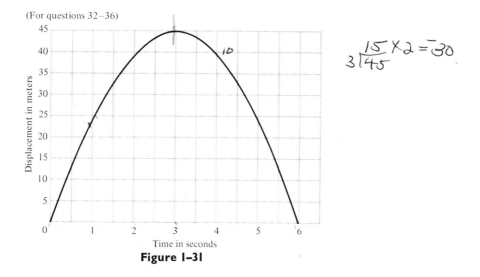

(For questions 32–36)

$$3\overline{)45}^{\frac{15}{45}} \times 2 = 30$$

Time in seconds

Figure I-31

Figure 1–32 shows the displacement of a car traveling in a straight line as a function of time. Answer the following questions:

37. At what time was the displacement zero? 15 sec.

38. When was the velocity zero? 10 sec.

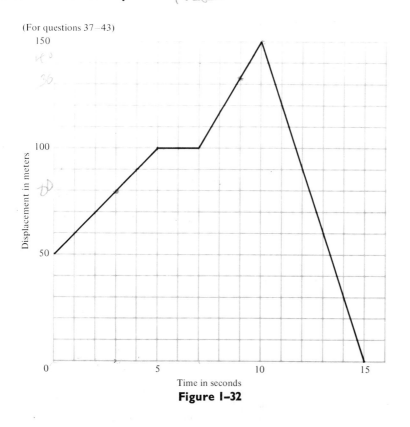

(For questions 37–43)

Time in seconds

Figure I-32

39. What was the maximum displacement?

40. What was the velocity at $t = 3$ seconds?

41. At what time(s) was the velocity negative?

42. What was the value of the negative velocity?

43. What was the velocity at $t = 9$ seconds?

A car is 70 meters in a straight line from point 0 and has a constant velocity of −4 m/sec.

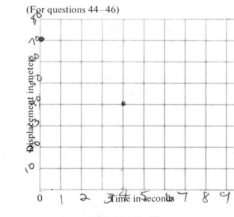

(For questions 44–46)

$$-4 = \frac{70}{t}$$

Figure I–33

44. Draw a displacement/time curve.

45. At what time will the car arrive at point 0?

46. At what time will the car be 30 meters from point 0?

FORCE, MOTION, AND ENERGY

ACCELERATION

Often the velocity of a body changes. For example, a car may start from rest and attain a velocity of 60 miles/hour north. We are interested not only in how much the velocity changes in direction or in magnitude (or both), but also in how much time is required to make the change. For example, suppose that a car is going 20 m/sec north and that it gains speed in a constant manner until it attains a velocity of 100 m/sec north. Let's

GRAPH 2-1

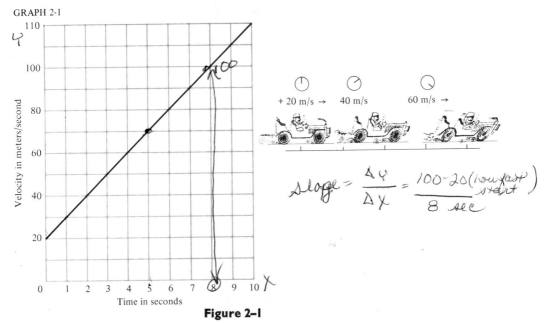

Figure 2–I

look at a plot (Graph 2–1) of the velocity as a function of time. The plot in this case is a straight line with constant slope, which is equal to:

$$\frac{(100 - 20) \text{ m/sec}}{(8 - 0) \text{ sec}} = \frac{10 \text{ m/sec}}{1 \text{ sec}} = 10 \text{ m/sec}^2.$$

The slope tells us the rate of change of the velocity. Acceleration is the rate of change of velocity with time.

Example 1:   What was the velocity at $t = 5$ seconds?

Answer: 70 m/sec.

Example 2:   What was the acceleration at $t = 5$ seconds?

Answer: 10 m/sec².

Example 3:   Is the acceleration constant?

Answer: Yes.

Acceleration is a vector quantity, since a vector (Δ velocity) divided by a scalar (Δt) is a vector.

(1) Constant acceleration $= \dfrac{\Delta \text{ velocity}}{\Delta \text{ time}}$

(plots straight line)

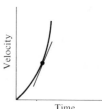

(2) Average acceleration $= \dfrac{\Delta \text{ velocity}}{\Delta \text{ time}}$

(uses points only)

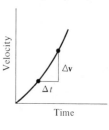

(3) Instantaneous acceleration $= \lim\limits_{\Delta t \to 0} \dfrac{\Delta \text{ velocity}}{\Delta \text{ time}}$

(slope at particular point)

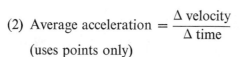

Figure 2–2

Acceleration is the slope of a velocity versus time curve, and we have three accelerations with some meaning, just as we had three velocities and three speeds. They are:

Constant acceleration tells us that the change in velocity is uniform with time. That is, if a car gains 20 m/sec north the first second, it will gain 20 m/sec north in every succeeding second. (Most problems in introductory physics books assume constant acceleration.)

Average acceleration only tells us the net rate of change of velocity over a given period of time.

Instantaneous acceleration is the slope of the velocity versus time curve at a specific time.

The direction of the acceleration is the direction of the change in velocity. For example, if a car is traveling in a straight line and is gaining velocity, then the acceleration is in the same direction as the motion of the car. If the car is slowing down, the acceleration is in the direction opposite to the motion of the car. (We sometimes use the term deceleration to indicate this.)

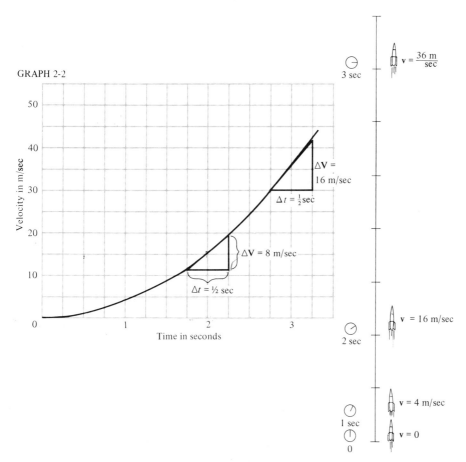

Figure 2–3

The following examples will illustrate these concepts of acceleration.

Example: A rocket is traveling in a straight line, as described by Graph 2–2.

(a) Is the acceleration constant?

(b) What is the velocity at $t = 2$ seconds?

(c) Calculate the average acceleration for the first two seconds.

(d) Calculate the instantaneous acceleration at $t = 2$ seconds.

(e) Calculate the instantaneous acceleration at $t = 3$ seconds.

Answers: (a) No, a constant acceleration would be a straight line.

(b) 16 m/sec.

(c) Average acceleration $= \Delta V/\Delta t = $ 16 m/sec/2 sec $= 8$ m/sec² (in direction of motion).

(d) Instantaneous acceleration = slope. Since slope is not constant, ΔV and Δt should be small in order to keep the curve between the points as straight as possible. Let $t = \frac{1}{2}$ sec. Then $V = 8$ m/sec, $\mathbf{a} = \lim_{\Delta t \to 0} \Delta V/\Delta t = 8$ m/sec/sec $= 16$ m/sec² (in direction of motion).

(e) Instantaneous acceleration $= 16$ m/sec/ $\frac{1}{2}$ sec $= 32$ m/sec² (in direction of motion).

Many times the acceleration is negative, which indicates that the change in velocity is negative, that the object is slowing down or gaining velocity in the direction opposite to that which we specify as being the positive direction. Let us look at an example.

Suppose that a proton (the nucleus of a hydrogen atom) is traveling in a straight line, as described by Figure 2–4. Let's describe the motion of the proton.

We will declare that velocities to the right are positive and velocities to the left are negative. Since the slope of the curve is constant, the acceleration is constant. The acceleration is negative, since the curve goes

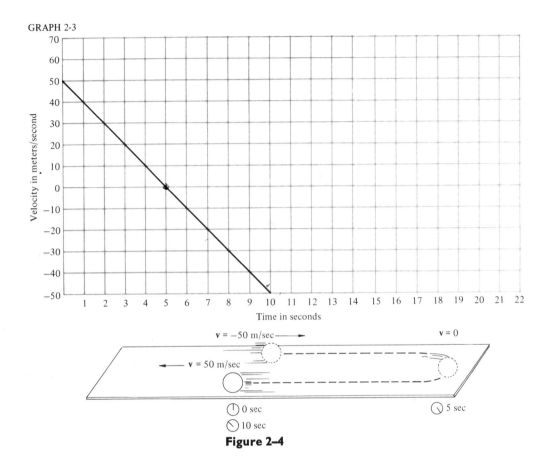

GRAPH 2-3

Figure 2-4

downhill, which tells us that the proton is gaining velocity in the direction opposite to the one we declared positive. The proton's original velocity is $+50$ m/sec. At the end of 5 seconds, it has lost all velocity in the positive direction and has stopped ($v = 0$). It now gains velocity in the opposite direction, and at $t = 10$ seconds the velocity of the proton is equal to the original velocity but is in the opposite direction; that is, $v = -50$ m/sec. Since the curve stops at $t = 10$ seconds, we cannot tell what happens after this time. Note that we can tell at what time the velocity was some specified value by reading directly from the graph. If someone wanted to know at what time the velocity was -25 m/sec, we would find it was at $t = 7.5$ seconds.*

Another example that is very important is the acceleration of an object near the Earth's surface. If we dropped an object from rest, a plot of velocity versus time would yield a downward curve (a positive velocity is upward; therefore velocity downward is negative).

---

* Graph 2-3 would not be sufficient for two-dimensional motion.

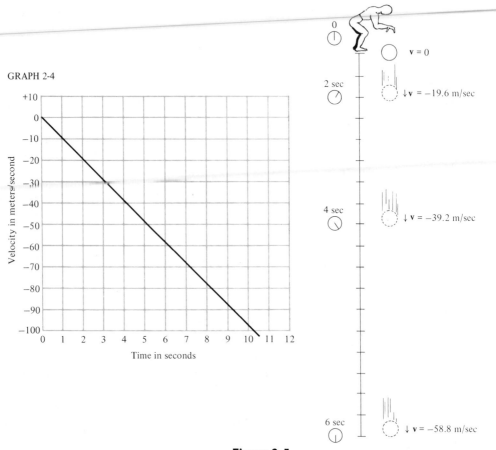

GRAPH 2-4

Figure 2–5

We find that the slope is constant, which tells us that the acceleration of an object toward Earth is constant near the Earth's surface. The value of the acceleration (the slope) is 9.8 m/sec² and is denoted by a special symbol, **g**.

$$\mathbf{g} = 9.8 \text{ meters per second per second toward earth.}$$

The value of **g** changes slightly over the surface of the Earth and changes with height above the Earth, but for problems involving heights less than 160 kilometers (around 100 miles) we will consider the value of **g** as constant.

THE AREA UNDER A CURVE

The concept of slope is very useful in expressing quantities that are ratios or rates. However, there are many quantities in physics that are products. We need the concept of area under a curve to handle products.

When we speak of "area under a curve," we are using the term "area" in a much broader sense than the number of square feet or square meters of surface. For example, if a train is traveling at a constant velocity of 20 meters per second north and we wish to know the displacement from the zero position* after 30 seconds, we could compute the displacement with a little algebra, since:

$$\text{velocity} = \frac{\Delta \text{ displacement}}{\Delta \text{ time}} \quad \text{or} \quad \Delta \text{ displacement} =$$

$$(\text{velocity})(\Delta \text{ time})$$

or

$$\Delta \text{ displacement} = (20 \text{ m/sec N})(30 \text{ sec}) = 600 \text{ m north.}$$

We can also compute the displacement by another method. Let's plot the curve of the velocity as a function of time, as in Graph 2–5 (Figure 2–6).

GRAPH 2-5

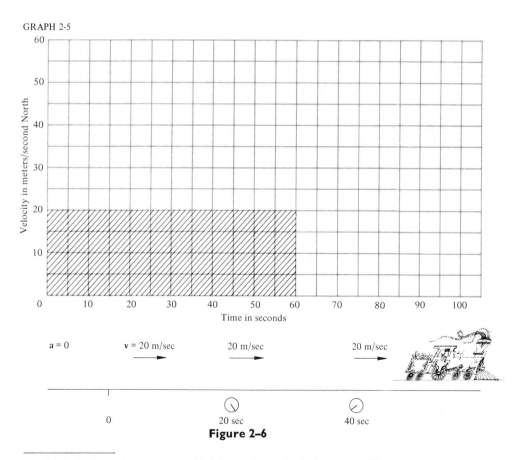

Figure 2–6

_____

* Displacement from a zero position is the change in displacement, ΔD.

The "area" under the curve $(v = 20$ m/sec$)$ to $(v = 0$ m/sec$)$ and between $(t = 0$ sec$)$ and $(t = 30$ sec$)$ is $(20$ m/sec north$)(30$ sec$) = 600$ m north. We obtained the same displacement by using the area under the curve as we did by algebra. Therefore, calculating the area under the curve is an alternate method of finding a product in this case. If all problems were this simple, algebra would be a sufficient method to find products. However, some problems cannot be solved by simple algebra.

Suppose that a plane does not have a constant velocity, but starts from rest and accelerates due north, as given in Graph 2–6. If we want to know the displacement from zero after 5 seconds, we cannot use simple algebra, because the velocity is changing every instant. What we can do is to break the 5-second time interval into 10 half-second intervals, find the displacement for each half-second time interval, and add all the displacements together to find the total displacement. The velocity still changes every 1/2 second by 5 m/sec.

It might seem wise to use a shorter interval, say 1/10 second, because the velocity changes only 1 m/sec every 1/10 second. Following this logic, we could decrease the time interval to 1/100 second or even to 1/1000 second so that the change in velocity would be even less. However, the number of calculations of displacements increases as the time interval decreases. For 5 seconds of elapsed time, a 1/2 second time interval requires 10 calculations of displacement, a 1/10 second time interval requires 50 calculations, a 1/100 second time interval requires 500 calculations, and a 1/1000 second time interval requires 5000 calculations. So, although our approximation for the true value of the displacement gets better as the time interval gets smaller, the number of calculations gets larger. Algebra is not a sufficient mathematical tool for this type of problem, since we would like the time intervals to be "infinitely" small in order to get the true value of the displacement.

Finding the displacement by the area under the curve is tantamount to dividing the elapsed time interval (Δt) into infinitely small time intervals, calculating the change in displacement for each infinitely small time interval, and adding all the changes in displacements together to find the total change in displacement.*

Let's calculate the displacement for the plane (Graph 2–6), using the area under the curve method. The area (the shaded portion) is a triangle, so the area is: 1/2 (base)(height) *or*

$$\Delta \text{ displacement} = \tfrac{1}{2}(\Delta t)(v)$$

$$= \tfrac{1}{2}(5 \text{ sec})(50 \text{ m/sec north}) = 125 \text{ m}$$

A very easy calculation! *Therefore, whenever the product of (quantity plotted along ordinate)(quantity plotted along abscissa) is needed, use the "area under the curve" method to calculate the needed product if the value of the quantity along the ordinate varies.*

---

* Since we must find area under curves from graphs, our calculations are somewhat limited. By using integral calculus, the exact area under curves can be found.

GRAPH 2-6

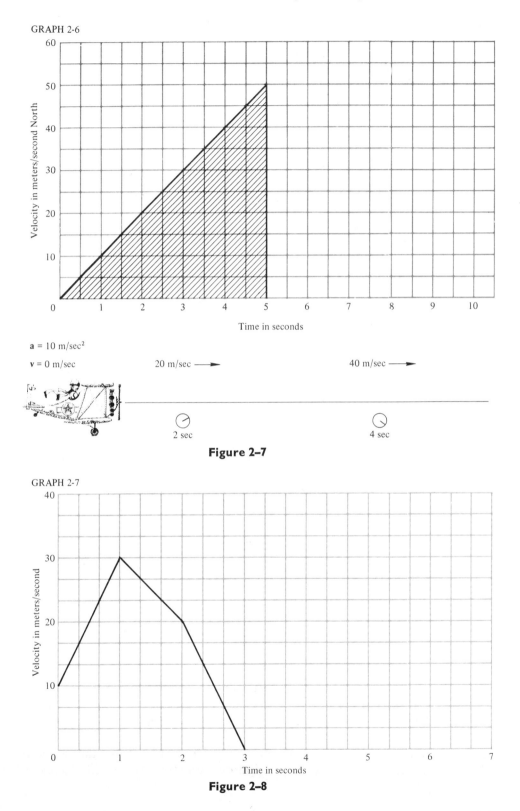

a = 10 m/sec²

v = 0 m/sec   20 m/sec ——→   40 m/sec ——→

2 sec   4 sec

Figure 2–7

GRAPH 2-7

Figure 2–8

Let's look at a more complicated example. Graph 2–7 shows the velocity of an object moving in a straight line as a function of the time. Let's say that the object is an alpha particle (a helium nucleus) traveling 10 m/sec. It accelerates uniformly to 30 m/sec in one second, then negatively accelerates from 30 m/sec to 20 m/sec in one second, and then negatively accelerates from 20 m/sec to 0 m/sec in one second. Find the total displacement. (Assume straight line motion.)

In this case, if we tried to find the displacement by simply multiplying (velocity)(time), we would be absolutely wrong, because the velocity changes every second. The method of finding the displacement by the area under the curve takes into account the changing velocity and gives a true value for the displacement. An easy method is to find the areas under the first, second, and third seconds separately and then add the three areas together.

that is, for the 1st second

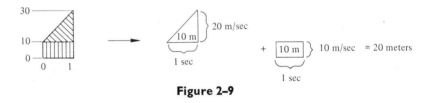

Figure 2–9

for the 2nd second

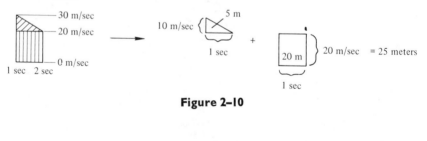

Figure 2–10

for the 3rd second

The total displacement is 20 + 25 + 10 = 55 meters.

Figure 2–11

The total displacement is $20 + 25 + 10 = 55$ m.

We could have found the area under the curve many other ways to get the same answer. Let's try still another example, which is illustrated by Graph 2–8.

Example: A brave hunter's car uniformly accelerates in a straight line from 20 m/sec to 30 m/sec in 20 seconds, after which it accelerates from 30 m/sec to 50 m/sec in 10 seconds.

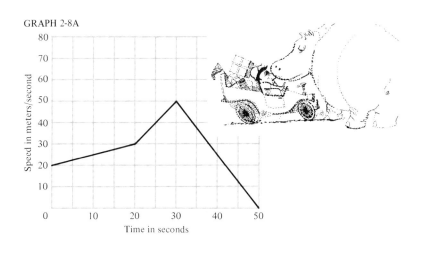

GRAPH 2-8A

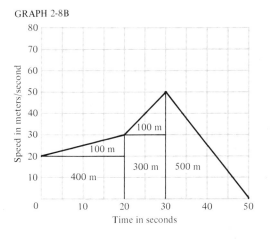

GRAPH 2-8B

Figure 2–12

The hunter then negatively accelerates uniformly to 0 m/sec in 20 seconds. Let's find out the following things about the hunter:

(a) How far did the hunter travel during the first 20 seconds?

(b) How far did the hunter travel from $t = 20$ to $t = 30$ seconds?

(c) How far did the hunter travel from $t = 30$ to $t = 50$ seconds (the distance it takes the hunter to stop from 50 m/sec)?

(d) What was the total distance traveled?

(e) What relation do you think there would be between the negative acceleration of the hunter's car and his blood pressure? (No answer provided.)

Answers: In order to find the distance, we must find the product of the (speed)(time) under the curve. Therefore, from $t = 0$ to $t = 20$ seconds, we must add the area of a triangle 20 seconds "wide" and 10 m/sec "high" to the area of a rectangle 20 seconds "wide" and 20 m/sec "high" (see Figure 2–12A).

(a) $\frac{1}{2}(20 \text{ sec})(10 \text{ m/sec}) + (20 \text{ sec})(20 \text{ m/sec})$
100 m + 400 m = 500 meters.

(b) From $t = 20$ to $t = 30$ seconds:
area of triangle = $[\frac{1}{2}(50 - 30) \text{ m/sec}][(30 - 20) \text{ sec}] = 100$ m
area of rectangle = $[(30 - 0) \text{ m/sec}][(30 - 20) \text{ sec}] = 300$ m.
100 m + 300 m = 400 m

(c) From $t = 30$ to $t = 50$ seconds:
area of triangle = $[\frac{1}{2}(50 - 0) \text{ m/sec}][(50 - 30) \text{ sec}] = 500$ m.

(d) Total distance = sum of distances = 500 m + 400 m + 500 m = 1400 m.

This problem might seem complicated, but with a little practice you will solve very interesting problems with a minimum of effort. Remember that the area under a velocity versus time curve gives the change in displacement; the area under a speed versus time curve gives the change in distance. You *must* use the area under the curve method to solve the problem if the ordinate changes.

We will now look at the concept of negative area. The area under a curve can be negative and does have physical meaning. For example, suppose that a ball rolls directly toward a wall with a constant velocity of 2 m/sec and rebounds with a velocity of -2 m/sec (the minus sign indicates that the velocity after rebound is in the direction opposite to that of the original velocity.) This is shown in Figure 2–13.

Notice that the ball hits the wall at $t = 2$ seconds. The displacement

GRAPH 2-9

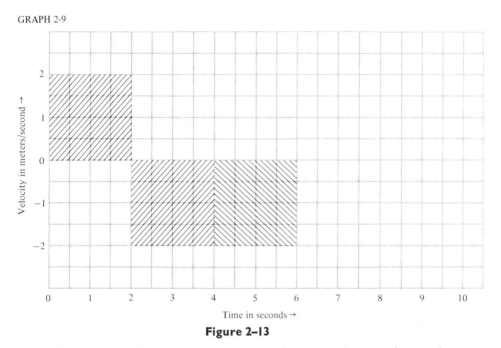

Figure 2–13

from the zero position at $t = $ zero seconds to $t = 2$ seconds can be computed by the positive area and is equal to (2 m/sec)(2 sec) = 4 meters. After 2 seconds, the velocity and the area are both negative. This indicates that the ball has started toward its original position.

To find displacement, we still add (algebraically) the areas under the curve.*

At $t = 3$ seconds, displacement = (2 m/sec)(2 sec) + (−2 m/sec) (1 sec) = 4 m − 2 m = 2 meters. The ball was 2 meters from the original position but is returning to the original position, since some area is negative.

At $t = 4$ seconds, displacement = (2 m/sec)(2 sec) + (−2 m/sec) (2 sec) = 4 m − 4 m = 0 meters. The ball has now returned to the original position.

At $t = 6$ seconds, displacement = (2 m/sec)(2 sec) + (−2 m/sec) (4 sec) = 4 m − 8 m = −4 m. The ball has passed the original position and has traveled 4 meters in the opposite direction.

If a positive area indicates a displacement in one direction, a negative area represents a displacement in the opposite direction.

at t = 2 seconds   0 —— 4 m ——▶

at t = 3 seconds   —— 4 m ——▶ + ◀— 2 m — = — 2 m —▶

at t = 4 seconds   —— 4 m ——▶ + ◀— 4 m — = 0 m

at t = 6 seconds   —— 4 m ——▶ + ◀———— −8 m ———— = ◀——— −4 m

---

* Area under the curve is the area calculated from the curve to the abscissa, regardless of whether the curve is above or below the abscissa.

Look at another example. Remember, the area under a velocity versus time curve gives the displacement, and the area of a curve is negative if it's below the origin and positive if it's above the origin.

Example: Figure 2–14 shows the results of an experiment in which a ball is rolled straight up a hill with an initial velocity of 20 m/sec. Find the displacement at the end of (a) 4 seconds, (b) 8 seconds.

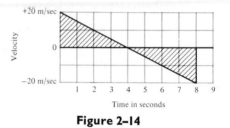

Figure 2–14

Answers: (a) At 4 seconds, area is ½(20 m/sec)(4 sec) = +40 m. (b) At 8 seconds, area is:

$$\frac{1}{2}(20 \text{ m/sec})(4 \text{ sec}) = +40 \text{ m}$$

$$\frac{1}{2}(20 \text{ m/sec})(4 \text{ sec}) = \frac{-40 \text{ m}}{0 \text{ m}}$$

At 8 seconds, the ball has rolled back down the hill to the starting point.

Let's take still another example. Two nuclear particles accelerate in opposite directions, as shown by Graph 2–10. How far apart are the particles in 8 milliseconds?

In this case, the accelerations of the particles are in opposite directions, so the slope of the velocity versus time curve of one is positive and the slope of the other is negative. The area under the curve of one is positive, indicating that the displacement is in one direction, and the area under the curve of the other one is negative, indicating a displacement in the opposite direction. The difference of the two displacements, which is what we are seeking, is the area between the curves, that is, (area above − area below).

---

* It is recommended that Experiment 2 in the lab manual be done at this time.

GRAPH 2-10

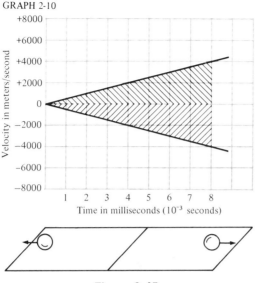

Figure 2–15

The area of triangle above the time axis is $\frac{1}{2}(4000)(8 \times 10^{-3}) = 16$ m. The area of triangle below the time axis is $\frac{1}{2}(-4000)(8 \times 10^{-3}) = -16$ m. We now find the difference, that is, (area above − area below) = 16 m − (−16 m) = +32 m. The two particles are 32 meters apart after 8 milliseconds.

NEWTON'S THREE LAWS OF MOTION

We have studied the description of motion in some detail. The description of motion is called kinematics. Now we study the far deeper question of why things move. Studying the cause of motion is that branch of physics called dynamics.

Sir Isaac Newton (1642–1727)—*English—Truly one of the greatest intellects of all time. He was not a precocious child, but, being an introvert, he applied himself diligently to study.*

The year he received his degree (1665), the plague closed the university for the next two years. During this period, Newton (age 23–25) discovered the theory of colors, the binomial theorem, differential calculus, and integral calculus, and conceived the concept of universal gravitation when he saw an apple fall from a tree.

Figure 2–16

In 1668, he invented the reflecting telescope, and in 1687 he published the Principia, perhaps the greatest scientific work ever written.

For all his accomplishments, Newton was a modest man, who commented, "I do not know what I may appear to the world, but to myself I seem to have been only like a boy playing on the seashore, and diverting myself in now and then finding a smoother pebble or a prettier shell than ordinary, whilst the great ocean of truth lay all undiscovered before me."

Sir Isaac Newton established the laws of motion. In order to understand the laws of motion, let's take an imaginary trip out in space, far from the influences of gravity. We will perform the following experiment. We take two chunks of *identical* material of the same size, each with a very light spring on one end.* We push the masses together in such a way that the springs are compressed (Figure 2–17). We release the masses, and the acceleration of chunk 1 and the acceleration of chunk 2 are equal and in opposite directions. We write: $\mathbf{a}_1 = -\mathbf{a}_2$ *or* $\mathbf{a}_2 = -\mathbf{a}_1$, where $\mathbf{a}_1$ is the acceleration of chunk 1 and $\mathbf{a}_2$ is the acceleration of chunk 2.

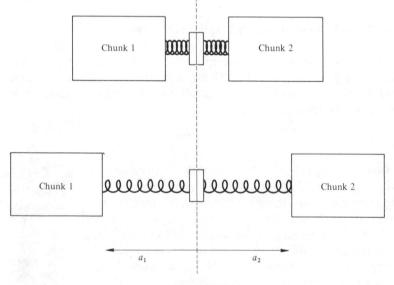

Figure 2–17

---

* The springs are so light that we can ignore their motion.

Next, we saw chunk 1 in half and repeat the experiment, using half of chunk 1 and all of chunk 2, as illustrated in Figure 18.

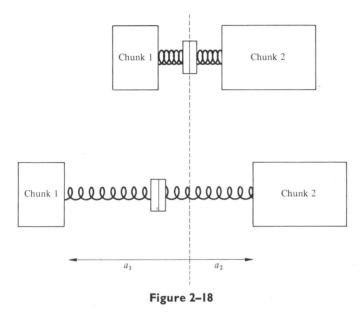

Figure 2–18

We release the masses and find that chunk 1 accelerates twice as fast as chunk 2, or $\mathbf{a}_1 = -2\mathbf{a}_2$. If the springs were elongated rather than compressed, the equation would be the same, except that $-\mathbf{a}_1 = 2\mathbf{a}_2$.

All we did was to saw chunk 1 in half, so there must be a relationship between the acceleration of an object and the amount of material in the object. The smaller an object, the greater its acceleration. If we assign a property m_1 to chunk 1 and a property m_2 to chunk 2, we can write this as: $m_1\mathbf{a}_1 = -m_2\mathbf{a}_2$. Now, if we declare that property m_2 of chunk 2 a fundamental quantity of mass with a standard unit (say, 1 kilogram), then all other masses can be compared to chunk 2 by repeating the preceding experiment. For example, if $\mathbf{a}_1 = 5$ m/sec and $\mathbf{a}_2 = -100$ m/sec, and chunk 2 is by definition the standard kilogram, then:

$$m_1\mathbf{a}_1 = -m_2\mathbf{a}_2$$

$$m_1 (5 \text{ m/sec}^2) = -(1 \text{ kilogram})(-100 \text{ m/sec}^2)$$

$$m_1 = 20 \text{ kilograms.}$$

Now we ask ourselves the very basic question, why does m_1 accelerate? We know that m_1 interacts with m_2 and that both accelerate, and we wonder if we could make m_1 accelerate without interacting with and causing the acceleration of some other mass. No matter how devious an experiment we perform, we cannot make m_1 accelerate without interacting with and causing the acceleration of some other mass. The results of these experiments are expressed in Newton's third law of motion.

Newton's Third Law.   *For every action, there is an equal but opposite reaction.*

We can express this idea more explicitly if we define action and reaction. The *action on m_1* (due to the interaction with m_2) is a quantity called *force*. That is, if $m_1\mathbf{a}_1 = -m_2\mathbf{a}_2$, then $\mathbf{F}_1 = m_1\mathbf{a}_1$ (where $\mathbf{F}_1$ is the force on m_1 due to $m_2\mathbf{a}_2$ and $\mathbf{F}_2 = m_2\mathbf{a}_2$ (where $\mathbf{F}_2$ is the force on m_2 due to $m_1\mathbf{a}_1$).

Therefore, Newton's third law can also be expressed: *For every force, there is an equal but opposite force.*

Newton's Second Law.   Newton's second law is a special case of the third law and is used when we are interested only in the motion of a particular object. If we take the formula $m_1\mathbf{a}_1 = -m_2\mathbf{a}_2$ and define the reaction of $m_2\mathbf{a}_2$ as the "unbalanced" *force on m_1*, we can write $\mathbf{F}_1 = m_1\mathbf{a}_1$. Expressed in words, *Newton's second law says that when an unbalanced force acts on a body, the body will accelerate in the direction of the force and that the magnitude of the acceleration will be proportional to the force and inversely proportional to the mass of the body.*

Newton's second law actually defines what a force is: namely, it is that quantity that will cause mass to accelerate. We have three different units of force as defined by Newton's second law, but we will discuss only the mks unit, the newton. One newton is the unbalanced force required to cause a mass of one kilogram to accelerate one meter per second per second in the direction of the force.

FORCE	=	(MASS)	×	(ACCELERATION)
1 newton or 1 kg m/sec²	=	1 kilogram	×	1 meter/sec²

Force is a vector quantity, since a scalar (mass) times a vector (the acceleration) is a vector. In using the formula $\mathbf{F} = ma$ to solve problems, we must abide by the rules. In working problems, we must be very careful to have the proper units and to consider only those forces that are acting on the accelerated object. If mass is in kilograms, and acceleration is in meters per second per second, then the unit of force will be the newton.

There are usually many forces acting on a mass, such as the gravitational force holding a mass to the surface and an equal force of the surface pushing back on the mass. Also, there are frictional forces pushing against the motion of the mass. By an unbalanced force, we mean the vector sum of all forces, which would be the force "available" to change the velocity of the object. An unbalanced force would then cause the mass to accelerate as though it were a single force acting against the mass in the absence of all other forces. It is tantamount to having the mass floating in space and then applying a force that is equal to the unbalanced force.

Example 1:   A force of 300 N gives a body an acceleration of 10 m/sec².
What is the mass of the body?

Answer: $\mathbf{F} = 300$ N; $\mathbf{F} = ma$

$m = ?$ 300 N $= m(10$ m/sec$)$

$\mathbf{a} = 10$ m/sec² $m = \dfrac{300 \text{ N}}{10 \text{ m/sec}^2} = 30$ kilograms.

Example 2:   Graph 2–11 shows a force pulling a 100 kg mass due north
as a function of time. What is the acceleration at $t = 5$
seconds and at $t = 10$ seconds?

Figure 2–19

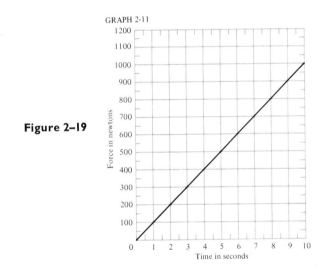

GRAPH 2-11

Answer: Reading from the graph at $t = 5$, the force is
500 newtons. Therefore:

$\mathbf{F} = ma$

500 N $= (100$ kg$)(\mathbf{a})$

5 m/sec² $= \mathbf{a}$ (the acceleration is due north because the
force is due north).

At $t = 10$ seconds, the force is 1000 newtons.

$\mathbf{F} = ma$ $\mathbf{F} = 1000$ newtons

$m = 100$ kilograms

$\mathbf{a} = ?$

1000 N $= 100 \cdot \mathbf{a}$

10 m/sec² $= \mathbf{a}$ (again, the direction is due north).

Newton's First Law. Newton's first law of motion is actually a special case of the second law, namely, what happens when no force acts on an object. Put another way, since $\mathbf{F} = ma$ and the force acting on an object is zero, what will happen if the mass is not zero? Obviously, the acceleration will be zero. Newton's law states the fact of zero acceleration in the following manner: *If no external force acts on a body, a body at rest will remain at rest and a body in motion will remain in motion.*

Figure 2–20

THE CONSERVATION OF MOMENTUM

From Newton's third law, we can derive the very important law of conservation of linear momentum. *Momentum is defined as the product of mass and velocity:* Momentum = (mass)(velocity). Momentum is a vector by our rules concerning vectors; the direction of the vector is the direction of the velocity. We can show how this law was derived for two particles of constant mass.

Recall that Newton's third law could be stated $m_1\mathbf{a}_1 = -m_2\mathbf{a}_2$, which can also be written

$$m_1\left(\frac{\Delta \mathbf{V}_1}{\Delta t_1}\right) = -m_2\left(\frac{\Delta \mathbf{V}_2}{\Delta t_2}\right),$$

since $\Delta \mathbf{V}/\Delta t$ is acceleration by definition. In any reaction, $\Delta t_1 = \Delta t_2$; that is, chunk 1 will push on chunk 2 only as long as chunk 2 pushes on chunk 1, so $m_1\Delta \mathbf{V}_1 = -m_2\Delta \mathbf{V}_2$ or $m_1\Delta \mathbf{V}_1 + m_2\Delta \mathbf{V}_2 = 0$ [conservation of momentum].

This law holds for any number of particles. That is,

$$m_1 \Delta \mathbf{V}_1 + m_2 \Delta \mathbf{V}_2 + m_3 \Delta \mathbf{V}_3 + m_4 \Delta \mathbf{V}_4 \cdots m_n \Delta \mathbf{V}_n = 0.$$

The law states that the *change* in momentum is equal to zero or that the total momentum is equal to some constant.

A System. The term "system" is often used when we want to focus our attention upon one group of molecules, particles, or other objects. We call the group of objects in which we are interested *the system* and all other objects in the universe the *environment. An isolated system is one in which there is no interaction with the environment.* A system's total momentum can only be changed by external forces, that is, forces caused by something in the environment. Since all internal forces are equal in magnitude but of opposite direction, all forces originating in the system will cancel each other. Any conservation law in a system is welcomed by a physicist studying the system because every aspect of a system that remains unchanged gives more insight into a workable model for the system.

The conservation of momentum tells us that if we vectorially add all the momentum of all particles in an isolated system, we would get zero if the center of mass* of the system were at rest and a constant if the center of mass were moving with a velocity $\mathbf{V}_1$. If the system exploded from internal forces, we would still get zero if the center of mass were at rest, or would get the same constant if the center of mass were moving with the same velocity $\mathbf{V}_1$. In short, whatever happens within a system, it cannot change the momentum of the system. The law of conservation of momentum is a powerful method used to solve problems of interaction of two or more bodies.

THE ROCKET

If you understand the law of conservation of momentum, you can understand how a rocket works. A rocket does *not* push against the atmosphere to propel itself. A rocket is composed of two primary parts: chunk 1, which consists of all the machinery, astronauts, food, and instruments we wish to put into space, and chunk 2, which consists of the fuel, oxygen, and everything else we are going to throw away. It is absolutely necessary to have chunk 2 because we must have something to push against.

To simplify the situation, let's assume that we have a rather primitive type of spacecraft of mass 1001 kilograms far out in space, where no forces are acting on the craft. Let m_2 be the mass of the spacecraft at any time and let m_1 be that part of chunk 2 that we are going to throw away.

---

* The center of mass is the point around which all particles of a rotating system would rotate if allowed to move freely.

We throw out a 1 kg mass with a speed of 20 m/sec and we declare the position where we throw it out the zero position (Figure 2–21).

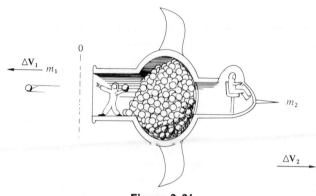

Figure 2–21

$$(m_1 \Delta V_1) = -(m_2 \Delta V_2), \text{ since mass is constant.}$$
$$(1 \text{ kg})(20 \text{ m/sec}) = -(1000 \text{ kg})(\Delta V_2)$$
$$\Delta V_2 = -2 \times 10^{-2} \text{ m/sec.}$$

The spacecraft changed its velocity by -2×10^{-2} m/sec relative to the position in space where we threw out the 1 kilogram mass, going in the opposite direction from the 1 kg mass. The spacecraft is now 1 kg lighter. We throw away another 1 kg mass with a velocity of 20 m/sec. By conservation of momentum:

$$(m_1 \Delta V_1) = -(m_2 \Delta V_2), \text{ since mass does not change}$$
$$\text{over the time interval.}$$
$$(1 \text{ kg})(20 \text{ m/sec}) = -(999)(\Delta V_2)$$
$$-2 \times 10^{-2} \text{ m/sec} \approx \Delta V_2.$$

Our velocity has changed again by slightly more than -2×10^{-2} m/sec relative to the point where we threw out the mass, or has changed -4×10^{-2} m/sec relative to our declared zero position.

If we continue to throw out 1 kg masses, the spacecraft will continue to lose mass until all of chunk 2 is gone. Every time we throw a 1 kg mass away, the change in the velocity increases slightly because the spacecraft is lighter than it was the preceding time. For example, by the time the spacecraft weighs 401 kg, throwing away a 1 kg mass means:

$$(m_1 \Delta V_1) = (m_2 \Delta V_2), \text{ since mass is constant.}$$
$$(1 \text{ kg})(20 \text{ m/sec}) = (400 \text{ kg})(\Delta V_2)$$
$$5 \times 10^{-2} \text{ m/sec} = \Delta V_2.$$

The change in the velocity is 2½ times what it was when we threw the first 1 kilogram mass away.

If we continue to throw out 1 kg masses, the rocket will continue to lose mass, and the change in velocity will increase each time. If the masses are thrown out often enough, the change in velocity will appear smooth, and we would interpret it as an increasing acceleration. In a typical rocket, the mass that is thrown out the rear (chunk 2) is hot gases, traveling with speeds of several thousand meters per second. The velocity of the exhaust and the amount of material thrown away each second are usually constant; therefore, the force on the rocket is constant. The increasing acceleration is due to the decreasing mass of the spacecraft.

The only way to increase the force (thrust) is to throw chunk 2 away at a faster rate or to increase the velocity of each part of chunk 2 we throw away. For example, had we thrown away the first kilogram of mass with a velocity of 2×10^5 m/sec instead of 20 m/sec, the change in the velocity of the rocket would have been:

$$(m_1 \Delta V_1) = -(m_2 \Delta V_2), \text{ since mass is constant.}$$
$$(1 \text{ kg})(2 \times 10^5 \text{ m/sec}) = -(1000 \text{ kg})(\Delta V)$$
$$-2 \times 10^2 \text{ m/sec} = \Delta V.$$

A very large velocity indeed! In order to increase the thrust, we have a choice of throwing away large amounts of mass at a relatively low velocity or small amounts of mass at a high velocity. Since it is not advantageous to carry large amounts of mass to throw away, we need a rocket engine that exhausts matter at a much faster rate than is possible with hot gases. Research is being done on this problem at the present time.

THE CONCEPTS OF WORK AND ENERGY

Work and energy are very important concepts in physics because they provide powerful, yet simple, methods for solving rather complicated problems. Work is defined as the scalar product of a force and displacement* if the two vectors (force and displacement) are parallel. If the two vectors are not parallel, we compute work with only the component of the force that is parallel to the displacement. All this information can be expressed in the formula:

<u>Work</u> = (displacement)(component of force parallel to displacement).

---

* The product of two vectors can be either a vector or a scalar. In this case it is a scalar.

Example 1:   Force = 50 newtons east

$\xrightarrow{\hspace{2cm}}$

Displacement = 10 meters east

$\xrightarrow{\hspace{2cm}}$

work = (50 N) east (10 m) east = 500 Nm

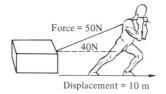

Displacement = 10 meters

Displacement = 10 m

Figure 2–22

Example 2:

Force
50 N

=
30 N
north

40 N east

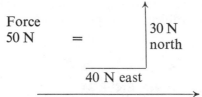

Displacement = 10 m east

Force = 50 newtons, but component parallel to displacement is 40 N, displacement = 10 meters east, therefore
W = (40 N) east (10 M) east = 400 Nm. (Note that work
is a scalar quantity, so there is no direction associated
with it.)

The mks unit of work is the newton meter, or joule.

Work = (force)(displacement)
joules = newtons   meters

The maximum amount of work is done on a system when the displacement and the force are parallel; the minimum amount of work (no work at all) is done when the displacement and the force are perpendicular to each other, as when a satellite is in a perfect circular orbit (Figure 2–23). The force on the satellite is always perpendicular to the displacement, so the component of the force parallel to the displacement is equal to zero. Therefore, the work done by or on the satellite is equal to zero when the satellite is in a circular orbit.

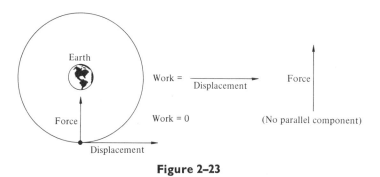

Figure 2–23

Energy. When work is done on a system by some outside or external force, the system must change in some manner. The system will either change its state of motion or change internally in some fashion. In the event that the system changes only its motion, we say that the system's _kinetic energy_ has changed. The amount of kinetic energy of any system can be calculated by the formula:

$$\boxed{\text{Kinetic energy} = \tfrac{1}{2}mv^2}$$

where m is the mass of the system and v is the speed of the center of mass.* For example, suppose that we calculate the kinetic energy of a 200 kilogram object traveling 20 m/sec.

$$KE = \tfrac{1}{2}mv^2$$

$$KE = \tfrac{1}{2}(200 \text{ kg})(20 \text{ m/sec})^2 = 4 \times 10^4 \text{ joules.}$$

On the other hand, we can perform work on a system without causing the system to increase its speed or to emit the energy in some other form. In that case, we say that the _internal energy_ of the system has increased. If the increase of the internal energy can be obtained from the

---

* This is assuming that the body is rigid and is not rotating. For a rotating system, the total kinetic energy is the sum of all the kinetic energies of all the different particles constituting the system.

system in some manner, we say that the system has increased its potential energy. We have actually stored energy in the system by changing the internal configuration in some way. The following are examples of systems with potential energy:

a compressed spring

a mass lifted above the floor

a lump of coal

an atom in an excited state

an atom capable of fission

two or more atoms capable of fusion

As you can see, potential energy arises from the position, condition, or state of being of a system. Any system in which the work put into the system changes the configuration, and in which the work put in can be regained from the system under ideal conditions, always has a change in potential energy associated with any change in the configuration. Such a system is called a *conservative system*.

If we work *on* the system, the change in potential energy is considered positive; if work is done *by* the system, the change in potential energy is considered negative.

To define zero potential energy, we arbitrarily pick a given point of configuration of the system and call it zero potential energy. For example, a person standing on the Earth's surface usually declares the Earth's surface to be the position of zero potential energy. Any object above this point has positive potential energy with respect to the Earth's surface and any object below this point has negative potential energy with respect to the Earth's surface. What we do concerning potential energy is the following: (1) We proclaim any convenient position (or configuration) as zero potential energy; (2) if work must be done on a system to get it to the declared zero position, the system has negative energy; (3) if work can be obtained from the system in getting it to the declared zero position, the system has positive potential energy.

An observer on the Earth looking upward at some mass sees it as having positive potential (Earth surface = 0 PE), whereas to an observer in space at zero potential energy everyone on Earth is in the bottom of a well.

Since there are many types of systems, there are many expressions for potential energy, such as potential energy of a mass near the Earth's surface, potential energy of a mass far from the Earth's surface, potential energy of a spring, and potential energy of electric charges. The change in potential energy for a particle can be calculated by finding the area under a force versus displacement curve (force and displacement are parallel). If there is more than one particle in the system, the potential energy change of each particle can be computed separately, and then all potential energy changes can be added together algebraically to find the total change in the potential energy of the system. Any system of particles must have conservative forces acting on the particles if there is to be an expression for the potential energy. A force is conservative if the work done on the particle by the force from point A to point B is equal to minus the work done from point B to point A. Or, put another way, the work done by the force on the particle is zero through a round trip. Any force that does not satisfy this condition is a non-conservative force.

Example: Find the change in gravitational potential energy of a mass that is lifted vertically above the Earth near the Earth's surface. Assume that the Earth does not move.*

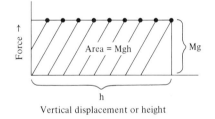

Vertical displacement or height

The change in potential energy is the product of the weight and the change in the height above the earth.

Figure 2–24

Answer: The force of gravity does not vary significantly near the surface of the Earth, so the force can be considered constant. The force of the Earth on the mass is by definition the weight, and would be equal to $\mathbf{F} = m\mathbf{a}$ or $\mathbf{F} = m\mathbf{g}$, since acceleration of gravity is denoted by $\mathbf{g}$. $\mathbf{F} = m\mathbf{g}$, $\mathbf{F}$ = weight, m = mass, $\mathbf{g}$ = acceleration = 9.8 m/sec², and h = displacement above Earth surface or height: $PE = m\mathbf{g} \cdot h.$

Examples: Graph 2–12 (Figure 2–25) shows the force on a spring as a function of the displacement when force and displacement are parallel.

(a) What was the force when the displacement was 3 meters? 7 meters? 12 meters?

---

* In reality, the Earth and mass above it constitute a system in which the mass falls toward the Earth and the earth toward the mass. The Earth is so massive it can be considered not to move.

GRAPH 2-12

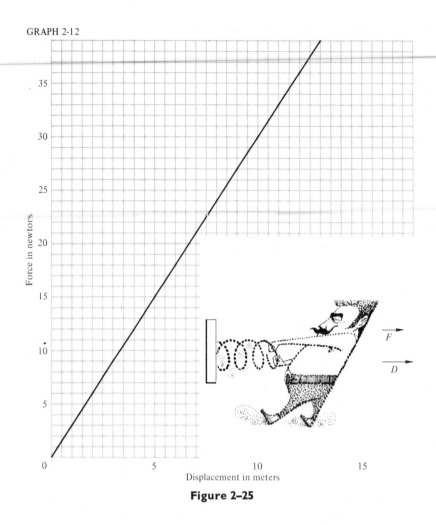

Figure 2–25

(b) Assuming that the force is conservative, what is the potential energy at 10 meters? Assume that the potential energy is zero when the force on the spring is zero.

(c) Plot the potential energy of this force as a function of the displacement.

Answers: (a) the height of the ordinate (which is the value of the force for any displacement) at 3 meters is 9 newtons, at 7 meters it's 21 newtons, and at 12 meters it's 36 newtons.

(b) Energy = (force)(displacement) where the two vectors are parallel, so we want a product function. To find this product, we want the area under the force versus displacement curve from zero to 10 meters. Since the area is a triangle, potential energy = ½(30 newtons)(10 meters) = 150 joules.

(c) In order to plot the potential energy curve, we need to know the potential energy as a function of displacement for several points. Therefore, we arbitrarily pick several displacements, solve for the potential energy as we did for question 2, and then plot the points. For example, suppose that we pick points 2, 4, 6, 8, 10, 12, and 14 for our points of displacement. The energy can be computed as follows:

TABLE 2-I

DISPLACEMENT (in meters)	FORCE (in newtons)	AREA UNDER CURVE OR POTENTIAL ENERGY (in joules)
2	6	6
4	12	24
6	18	54
8	24	96
10	30	150
12	36	216
14	42	294

We then plot the computed points of potential energy as a function of the displacement, as shown in Figure 2–26. Graph 2–13 is a parabola, and it is interesting to note that if the force versus the displacement plots a straight line, then the product function plots a parabola.* This would be true for any other functions. For example, if the velocity versus the time curve plots a straight line (slope not zero), then the product of the velocity time curve (which is the displacement) should plot a parabola.

GRAPH 2-13

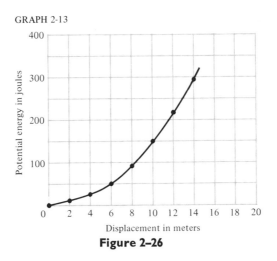

Figure 2–26

---

* Since the potential energy is positive, we are working on the spring, thus increasing the potential energy.

Binding Energy.   Imagine that you are an observer far out in space looking at Earth. As we stated previously, if you called your position the position of zero potential energy, all people on Earth would be in a "potential energy well," meaning that each person's potential energy would be negative relative to your position. If you calculate the magnitude of the work necessary to get someone out of the potential well, you have calculated the energy by which he was bound to the Earth as well. This energy is called binding energy and is always negative. If you calculated all the binding energy for every person, you would have the total binding energy of the (Earth-people) system. In many systems, such as the atom and the nucleus, we use the concept of binding energy for particular particles and total binding energy for the system.

If only conservative forces act within a system, any change in the kinetic energy of a particle in the system will always be accompanied by an opposite change in the potential energy. For example, an object falling toward Earth loses potential energy, but the speed of the object increases, so that the loss in potential energy is just equal to the gain in kinetic energy (if we neglect air resistance). We can write this as:

change in potential energy = change in kinetic energy.

Putting a minus sign to indicate loss of potential energy, we have:

$$-\Delta PE = \Delta KE \quad \text{or} \quad \Delta PE + \Delta KE = 0$$

or

$$PE - PE_0 + KE - KE_0 = 0$$

$$PE + KE = PE_0 + KE_0$$

This equation is known as the law of conservation of energy for conservative forces. It tells us that the total energy of an isolated system is constant.

Now imagine that you are an observer on Earth, making your measurement from the Earth's surface, and that you observe an object traveling at speed v_0 at height h_0. A little later you observe the object traveling at speed v at height h (Figure 2–27A). You would write the conservation law: $\overbrace{Mgh_0 - Mgh}^{\text{loss in PE}} = \overbrace{\tfrac{1}{2}mv^2 - \tfrac{1}{2}mv_0{}^2}^{\text{gain in KE}}$.

How would an observer out in space interpret the same event? If he made his measurements from his observation point, he would write:

$$\overbrace{mgd_0 - Mgd}^{\text{loss in PE}} = \overbrace{\tfrac{1}{2}ms^2 - \tfrac{1}{2}ms_0^2}^{\text{gain in KE}}$$

where d_0 is the original height

d is the final height

s_0 is the original speed

s is the final speed

m is the mass

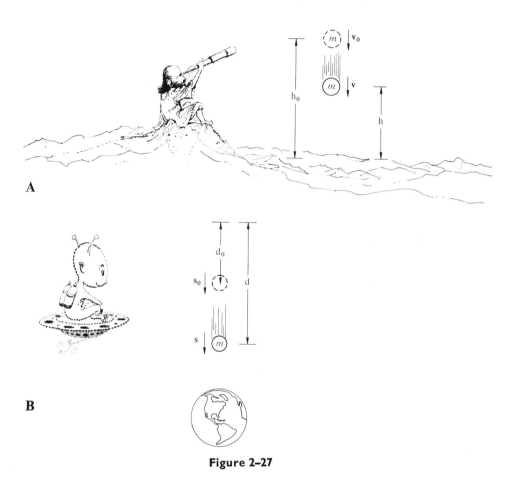

Figure 2–27

The observer in space would interpret the event exactly as the observer on Earth did (Figure 2–27B). One of the very important aspects of a conservation law is that two observers in two reference frames will agree that the same conservation laws would apply to the same event.

What is usually done is that an observer assigns 0 KE to a particle that is moving with a speed = 0 relative to his reference point and zero potential energy to a particle at his reference point. He then speaks of the kinetic energy and the potential energy of the particle. The law of conservation of mechanical energy for conservative forces can then be expressed as:

> Total energy = potential energy + kinetic energy = a constant.

Both total energy and potential energy can be positive or negative.*

Frequently, non-conservative forces also work on a particle in the system. If the forces are not small enough to be ignored, we must take

---

* It is recommended that Experiment 3 in the lab manual be done at this time.

the energy change caused by these forces into account. Experience has shown that when a non-conservative force (such as friction) works on an object, the energy loss will always show up in some other form (such as heat). Therefore, we can express the conservation of energy to include all forms of energy in the following way: Energy cannot be created or destroyed, but merely changed from one form to another. The total energy in any system remains constant if no work is gained or lost to the environment.

This statement is called the *principle of conservation of energy*. This principle has been formulated from experience and every apparent violation has been the topic of intensive research. In every case, the apparent violation has only resulted in more knowledge about and more faith in this principle.

The principle of conservation of energy provides a powerful yet simple method for solving rather complicated problems.

Example:　A boy slides down a hill that is 5.1 meters high. What is his speed at the bottom if frictional losses are ignored?

Answer:　Loss in potential energy = gain in kinetic energy

$$mgh = \tfrac{1}{2}mv^2$$

$$gh = \tfrac{1}{2}v^2$$

$$2\,gh = v^2$$

$$2(9.8 \text{ m/sec}^2)(5.1 \text{ m}) = v^2$$

$$10 \text{ m/sec} = v$$

Power. We need a measurement of how fast a machine can do work. This measurement is called "power" and is defined as:

$$\text{Power} = \frac{\text{total work done}}{\text{time interval to do work}} = \frac{\Delta W}{\Delta t}$$

If work is in joules and time is in seconds, the unit of power is the watt:

$$1 \text{ watt} = \frac{1 \text{ joule}}{1 \text{ sec}}$$

The watt was named in honor of James Watt, the Scottish inventor who made the first practical steam engine. A kilowatt, 1000 watts, is also a common unit of power. If the power as a function of time is known, it is an easy matter to compute the work done, since the work is the area under a power versus time curve. The power output of most motors is not constant, and the power rating is usually the maximum power the motor can produce without mechanical or electrical breakdown.

Example 1:   Graph 2–14 shows an experiment of the power output of a motor as a function of the time. What was the maximum power?

Answer: From the graph, $P = 300$ watts.

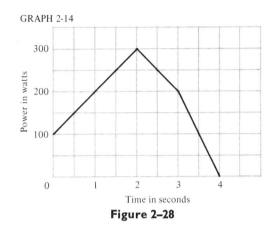

GRAPH 2-14

Figure 2–28

Example 2:   At what time was the power 100 watts?

Answer: From the graph, $t = 0$ sec and $t = 3.5$ sec.

Example 3:   How much work was done by the motor in 4 seconds?

Answer: Since $\Delta W = P \Delta t$ and power varies, we must use area under curve method. Taking the area under each second of time, we get $\Delta W = (150 + 250 + 250 + 100)$ joules $= 750$ joules.

EXERCISES

Figure 2–29 shows the results of an experiment in which a car rolls straight down a hill.

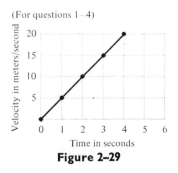

(For questions 1–4)

Figure 2–29

1. Is the acceleration constant?

2. What is the value of the acceleration at $t = 3$ seconds?

3. How far did the car travel in 5 seconds?

4. How far did the car travel during the fourth second (the fourth second is from $t = 3$ to $t = 4$)?

Figure 2–30 shows the velocity/time curve for a 50 kilogram rocket rising vertically.

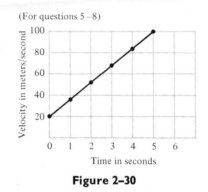

(For questions 5–8)

Figure 2–30

5. At what time was the velocity 90 m/sec?

6. What was the acceleration of the rocket?

7. What was the unbalanced force on the rocket?

8. If the rocket started from rest, how long was it in motion before the timer was started?

9. A force of 10 newtons acts northeast. What are the components due north and due east? (Compute, using a ruler.)

10. A heavy box is shoved 5 meters across a floor with a force of 500 newtons parallel to the floor. How much work was done?

Figure 2–31 shows the velocity of an object (the mass of the object is 2 kilograms) falling from rest in the Earth's gravitation field near the Earth's surface. Answer the following questions concerning the motion of the object.

11. Is the acceleration constant near the Earth's surface?

12. What is the value of the acceleration?

13. What was the velocity at the end of 4 seconds?

14. What was the force on the object?

(For questions 11–16)

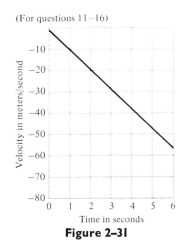

Figure 2–31

15. How far did the object fall in 5 seconds?

16. What was the kinetic energy of the object at the end of 5 seconds?

Figure 2–32 shows the force exerted on a spring as a function of the elongation (displacement) of the spring.

(For questions 17–19)

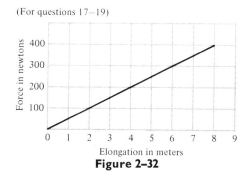

Figure 2–32

17. How much force was required to stretch the spring 4 meters?

18. How much work was done on the spring to stretch it 8 meters?

19. What was the potential energy of the spring at 8 meters if the unstretched spring had zero potential energy?

Figure 2–33 shows the velocity of a 2000 kilogram racing car traveling due north on level ground as a function of time. Answer the following questions concerning the motion.

20. What is the initial velocity?

21. When was the velocity 60 m/sec?

22. What was the acceleration at 2 seconds?

(For questions 20–30)

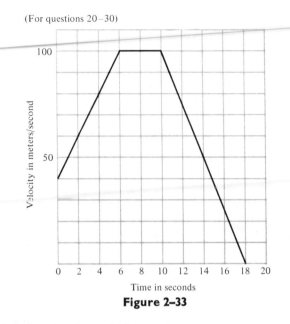

Figure 2–33

23. What was the acceleration at 14 seconds?

24. At what time was the acceleration 0?

25. At what time was the velocity 0?

26. How long was the object slowing down?

27. How far did the car travel while negatively accelerating?

28. What was the total displacement?

29. What was the unbalanced force on the car during the first 6 seconds?

30. What was the unbalanced force on the car during the eighth second?

31. What is the potential energy, relative to Earth, of a 2000 kilogram plane traveling 100 m/sec at an altitude of 1000 meters?

    (b) What is the total energy of the plane?

    (c) What is the minimal amount of work the motor had to do to get the plane off the ground and flying at this altitude?

32. If the motor in problem 31 did the work in 5 minutes, what is its power in watts?

33. A car traveling 20 m/sec has an acceleration of −4 m/sec² when the driver applies the brakes. The reaction time of the driver is 1.0 sec. Plot a graph and answer the following questions:

    (a) How long did it take the driver to stop?

    (b) How far did the car travel before the driver could apply the brakes?

(c) How far did the car travel while the driver was applying the brakes?

Figure 2–34 shows the results of an experiment in which a mass oscillates back and forth on a spring.

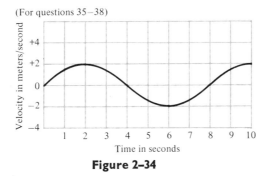

(For questions 35–38)

Figure 2–34

34. What is the acceleration at $t = 4$ seconds?

35. What is the maximum positive velocity?

36. When the velocity is a maximum, what is the acceleration?

37. What is the displacement (approximately) at $t = 4$ seconds?

38. What is the displacement (approximately) at $t = 8$ seconds?

39. Two atomic particles start from rest in opposite (straight line) directions. One particle has an acceleration of 10 m/sec² and the other of 20 m/sec². Draw a graph of the situation (for 6 seconds) and answer the following questions:

    (a) What was the velocity of the particles at the end of 4 seconds?

    (b) What was the distance between the particles at the end of 6 seconds?

    (c) What was the velocity of separation at $t = 5$ seconds?

40. Figure 2–35 shows the power output of a motor as a function of time. How much work did the motor do during the first 6 seconds?

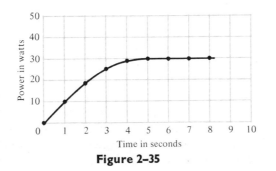

Figure 2–35

3

GRAVITATION AND THE EXPLORATION OF SPACE

Congratulations! If you have studied Chapter 1 and Chapter 2 diligently, you have mastered enough concepts and attained enough mathematical tools to look at nature through the eyes of a physicist. Let's look at the adventure of the exploration of outer space. First, we would like to have a model of the universe. This is quite a big order! Any model we make of the universe will certainly be crude, incomplete, and in many aspects probably incorrect—but it will be a model.

The search for a model of the universe has caused bitter controversy, threats, death, and one of the greatest triumphs of the human intellect. Let's see how it all began.

Some ancient astronomers, such as Aristarchus in Greece (around 320–250 B.C.), deduced what we fully accept today—that all the planets go around the sun—but this theory was rejected in favor of the theory of a Greek (or Egyptian) philosopher, Ptolemy, who advocated that the center of the universe was the Earth and that everything else revolved around it. This theory did a lot for man's ego, was not contradictory to the teachings of the Church, and was able to predict the motion of the planets well enough to agree with eye observation, although certain planets had to do a loop-the-loop (epicycles) to account for their motions in the sky. People accepted this model for 13 centuries, until Nicholas Copernicus (1473–1543), assuming a heliocentric (sun-centered) system, worked out the full mathematical details to predict the positions of the planets. Copernicus's model was still defective; he insisted upon perfect circles for orbits that still required epicycles. The Copernican system was far more mathematically elegant than the Ptolemaic, but charges of—

and punishment for—heresy were frequent and many times fatal, so Copernicus waited until he was at death's door (or so the story goes) before he published his theory.

Tycho Brahe (1546–1601) made observations of the planets over a period of years with an accuracy that no man has yet surpassed without the use of a telescope. Brahe gave his young assistant, Johann Kepler, his data. Kepler, who was a mathematician, discovered from the data the following beautiful and rather simple laws.

I. Each planet moves around the sun in an ellipse, with the sun at one focus. (The ellipse is a precise mathematical curve. To draw an ellipse, drive two nails in a board and make a loop out of a piece of string. Put the loop over the nails, each of which is a focus. If you put a pencil inside the loop and draw the string taut, you can trace the ellipse as you move the pencil one complete revolution.)

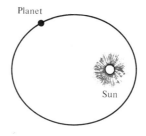

Planet

Sun

Figure 3–1

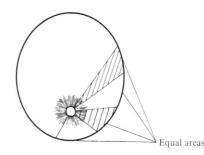

Equal areas

II. A vector from the sun to a planet sweeps out equal areas (under the curve) in equal times.

III. For all planets, the square of time for one revolution (the period) is proportional to the cube of its mean distance from the sun.

The Copernican theory with Kepler's modification provided a much better model than the Ptolemaic system because it was mathematically simple, it predicted with better accuracy, and it eliminated the need for epicycles. On the other hand, it relegated the abode of man (the Earth) to a far less distinguished place in the scheme of things and it was an empirical law—that is, it worked but no one knew why it worked.

While Kepler was describing *how* the heavenly bodies move, Galileo Galilei was studying *why* they move. The prevalent thought in those days was that the *natural* state of a body was to be at rest, and that anything that moved (even with constant speed) was being pushed by something. Everyone agreed that the planets—the word "planet" means wanderer—moved, and some medieval philosophers proposed that an angel was behind each planet, doing the pushing.

Galileo had a keen mind and took issue with the prevalent thoughts of that day. He discovered the essence of inertia, that the *natural* state of a body is to coast forever if it is in motion, or rest forever if it is not in motion.

Galileo also discovered the moons of Jupiter—proof positive that everything did not circle the Earth. Galileo attacked the Ptolemaic system in his masterpiece, *Dialogue on the Two Chief World Systems*, in which a character called Simplicio futilely attempts to argue the Ptolemaic system to an intelligent layman—with, of course, disastrous intellectual results. The Establishment of that day convinced the Pope that Galileo's Simplicio was a caricature of the Pope himself. Galileo was forced to recant his views or possibly be burned alive for advocating heretical ideas. He recanted his "heresy" that the Earth was not stationary, but as he rose from his knees, he muttered, ". . . and yet it moves."

Figure 3–2

Galileo Galilei (1564–1642)—*First great advocate of the experimental method in science. Galileo had a brilliant mind, a caustic wit, and a rebel personality—a combination of traits that caused him to make bitter enemies of influential people. His experiments with the accelerating masses (the legendary Leaning Tower of Pisa experiment) proved that the acceleration of a body near the Earth's surface is independent of its mass—a direct contradiction of Aristotle's teachings.*

He made one of the first telescopes and computed the sun's rotation by sun spots, and discovered mountains on the moon and the major satellites of Jupiter (the Galilean satellites). Galileo is known for his studies in projectile motion, strength of materials, and forces. He advocated the Copernican system.

The scientific revolution that began with Copernicus was soon to flourish under the genius of Newton. Using his three laws of motion (Chapter 2) and Kepler's laws, Newton deduced that every body in the universe attracts every other body with a force directly proportional to the mass and inversely proportional to the square of the distance between the centers of mass. It remained for the great experimentalist Cavendish in his famous experiment on "weighing the Earth" to establish the value of the constant G that makes the proportionality an equality. That is,

$$F = G \frac{m_1 m_2}{D^2}$$ where F is the magnitude of the force in newtons

m_1 is one mass in kilograms

m_2 is the other mass in kilograms

D is the minimum distance

between centers of mass

G is a constant and is equal to $6.66 \times 10^{-11} \frac{Nm^2}{kg^2}$.

Newton's law of universal gravitation applies to any two bodies: two lovers, or two particles, or a planet and the sun. The force is always attractive.

Example: What is the maximum force of gravitational attraction between two lovers, if we assume their masses of 50 and 100 kilograms to be centered in their hearts and the minimum distance between their hearts to be 1/10 meter

Answer: $F = G \frac{m_1 m_2}{D^2}$ where $F = $?

$$m_1 = 50 \text{ kg}$$

$$m_2 = 100 \text{ kg}$$

$$D = 10^{-1} \text{ m}$$

$$G = 6.66 \times 10^{-11} \frac{Nm^2}{kg^2}$$

$$F = (6.66 \times 10^{-11}) \frac{Nm^2}{kg^2} \times \frac{(50 \text{ kg})(100 \text{ kg})}{(0.1)(0.1)}$$

$$F = 3.33 \times 10^{-5} \text{ newtons.}$$

(Any lovers who are disappointed in this infinitesimally small force can at least take comfort in the fact that the force should not decrease with time.)

What a beautifully simple model for the universe: one relation for all bodies at all distances. Only a nagging question remains—if a gravitational force were acting on every body, why would all the planets not fall or spiral into the sun, since the sun exerts a tremendous force on each planet?

The answer must be that the force the sun exerts upon a planet or a planet exerts upon the sun does not add or subtract energy to the system over one revolution, because the addition or subtraction of energy would change the configuration, or the speed, or both. In our discussion about energy, we found that no energy was added to a body if the force on the body was perpendicular to the path along which the body was traveling. This would imply that all planets would be moving in circular orbits, which is contrary to the experimental data of Brahe and to the conclusions of Kepler. Although it is beyond the mathematical scope of this book, it can be proven that the total energy of a planet in an elliptical orbit is also constant; therefore, planets which move in either elliptical or circular orbits would go around the sun indefinitely.*

In an elliptical orbit, the kinetic and potential energy change at every instant, but the total energy remains constant (Figure 3–3). When a planet is at the farthest distance from the sun (called aphelion), it has maximal potential energy, minimal kinetic energy, and minimal speed. As the

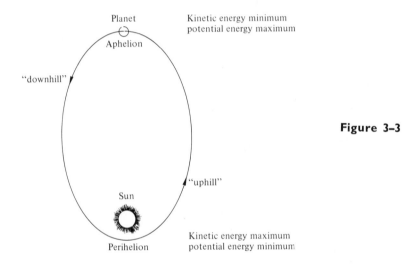

Figure 3–3

planet goes toward the sun, the force on the planet increases the speed of the planet, and the kinetic energy of the planet increases (the planet goes "downhill"). It gains kinetic energy and loses potential energy until it reaches the nearest point to the sun (called perihelion). At perihelion, it has minimal potential energy, maximal kinetic energy, and, of course, maximal speed. After perihelion, the force on the planet decreases the speed of the planet therefore, the planet loses kinetic energy and gains potential energy (the planet goes "uphill") until aphelion is reached again. At aphelion the cycle starts to repeat itself. The total energy remains constant and the planet coasts alternately "downhill" and "uphill" forever.

Circular Orbits. In a circular orbit, a satellite would remain a constant distance from the body to which it was bound. The speed, the total energy, the potential energy, and the kinetic energy are constant. Although no planets have circular orbits, some of the planets have nearly

---

* It is recommended that Experiment 4 in the lab manual be done at this time.

circular orbits. In this book, we will limit our mathematical descriptions to circular orbits because of the mathematical simplicity of circular orbits. We can still get an excellent idea of the magnitudes of such things as orbital speeds and binding energies. In order to study circular motion of a satellite, we must study the concept of centripetal force.

CENTRIPETAL FORCE

Whenever a body travels in a circle at constant speed, we know that the kinetic energy of the body does not increase and that the configuration does not change. Therefore, the forces acting on the body could not be

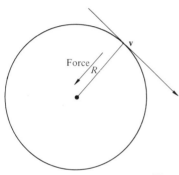

NOTE: The force on the body and the direction of the displacement of the body are always perpendicular. Therefore, at no time will work be given to or extracted from the body.

Figure 3–4

adding energy to the body. All the force is "used" to change the direction of the velocity, not its magnitude. We call such a force acting on the body a centripetal force, and the magnitude of the force is given by the relation:

$$F = \frac{mv^2}{R}$$ where F is the force

m is the mass of the body

R is the radius of the circle

v is the speed of the body

The direction of the force is along the radius.

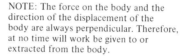

Example: What is the speed of a 1000 kilogram satellite just above the Earth's atmosphere?

Answer: A 1000 kilogram satellite just above the atmosphere would weigh almost the same as it would on Earth.

$$F = mg = (1000 \text{ kg})(9.8 \text{ m/sec}^2) = 9800 \text{ newtons.}$$

The radius of the Earth is 6.4×10^6 meters. The radius of the circle can be considered to be the Earth's radius, so:

$$F = \frac{mv^2}{R} \quad \text{where } F = 9800 \text{ newtons}$$
$$m = 1000 \text{ kg}$$
$$v = ?$$
$$R = 6.4 \times 10^6 \text{ m}$$

$$9800 \, N = \frac{(1000 \text{ kg})(v^2)}{6.4 \times 10^6}$$

$$v^2 = \frac{(9800 \, N)(6.4 \times 10^6 \text{ m})}{1000 \text{ kg}}$$

$$v = 8 \times 10^3 \text{ m/sec.}$$

The gravitational force between two bodies is the agent that supplies the centripetal force. If it were not for this force, a planet would go in a straight line forever, according to Newton's first law. It is the centripetal force exerted on each planet by the sun that makes the planet go around the sun. If we take body 1 of mass m_1 and body 2 of mass m_2 and let m_1 be large* and m_2 be small, and let m_2 circle m_1 with a speed v, then (Figure 3-5):

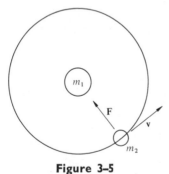

Gravitational force = centripetal force

$$G\frac{m_1 m_2}{R^2} = \frac{m_2 v^2}{R}$$

$$\boxed{v^2 = \frac{Gm_1}{R}} \quad \checkmark$$

Figure 3-5

This formula tells us that the orbital speed of any satellite in a circular orbit depends upon the radius of the circle, the mass of the body it circles, and the gravitational constant. This means that we cannot pick both our speed and our radius. Satellites close to the Earth must go faster than satellites farther away from the Earth, just as planets close to the sun go faster than planets farther away from the sun. Table 3-1 shows the

* Assume that m_1 is so large that it does not move. If m_1 is not massive, the bodies revolve around their common center of mass.

TABLE 3–1   THE SOLAR SYSTEM.

THE SOLAR SYSTEM

Planet	Mercury	Venus	Earth	Mars	Jupiter	Saturn	Uranus	Neptune	Pluto
Distance from Sun, 10^6 km	58	108	149	228	778	1426	2869	4495	5900
Velocity of Escape, km/sec	3.6	10.2	11.2	5.0	60	36	21	23	11?
Orbital Speed (km/sec)	48×10^3	35×10^3	30×10^3	24×10^3	13×10^3	9.6×10^3	6.8×10^3	5.4×10^3	4.7×10^3
Surface Gravity m/sec^2	2.6	8.3	9.8	3.6	26	11	9.0	14	?
Mean Diameter, km	5,000	12,400	12,742	6,870	139,760	115,100	51,000	50,000	12,700?

approximate speeds of planets in their orbits, as well as other pertinent data.

Note that Mercury has the fastest speed around its orbit and Pluto the slowest speed. It is surprising that a kilogram of mass in Pluto's orbit (but far away from Pluto) has more total energy than a kilogram of mass near Mercury's orbit. How can this be?

Imagine that you are an observer at the outer edge of the solar system, where you are so far away from the sun that the gravitational force is zero for all practical purposes. You declare your position to be the point of zero potential energy. If you threw a 1 kilogram ball toward the sun with an initial speed v but in such a way that the ball would miss the sun, the trajectory would be crudely as shown in Figure 3–6. As the ball left your

Figure 3–6

hand, the total energy of the ball-sun system would be positive, because you threw it with some initial speed. As the ball fell toward the sun, the potential energy of the ball would decrease, but the kinetic energy would increase by the same amount. Therefore, the total energy would be constant—and positive—and the ball would have too much energy to be bound to the sun. The centripetal force on the ball would turn the ball, but the ball would continue on into space after bypassing the sun, never to return.

Now let us drop down into the gravitational field of the sun, extracting work from our 1 kilogram ball as we go. For example, let's go to a position where the ball has lost 200 joules of potential energy to the environment relative to our zero potential energy point. The ball now has a total energy of −200 joules, since it lost the energy to the environment and it is not moving.

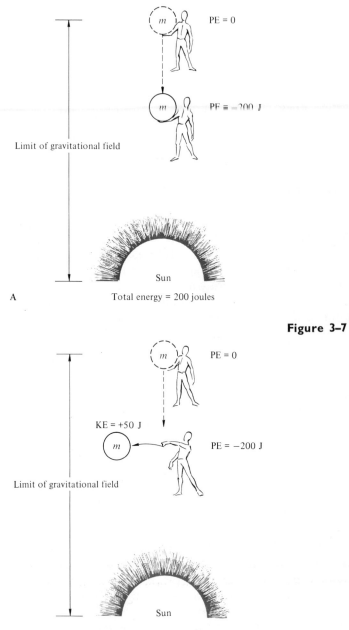

PE = 0

PE = −200 J

Limit of gravitational field

Sun

A   Total energy = 200 joules

Figure 3–7

PE = 0

KE = +50 J

PE = −200 J

Limit of gravitational field

Sun

Total energy = −150 J

B

We now throw our 1 kilogram ball with a speed such that the kinetic energy is less than 200 joules. Assume that we throw it with a speed of 10 m/sec, so that the kinetic energy is 50 joules. The total energy of the ball is now:

$$\text{PE} + \text{KE} = \text{total energy} \quad or \quad -200\text{ J} + 50\text{ J} = -150\text{ J}.$$

The ball has a binding energy of 150 J. The ball is bound to the sun, and it will have a trajectory something like the one shown in Figure 3–8. The

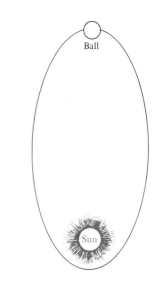

Figure 3–8

ball-sun system is now a bound system, and the ball will revolve around the sun forever unless the ball somehow gains energy from the environment.

For a given circular orbit, there is a permissible speed as given by the relation $v^2 = Gm_2/R$. To get closer to the sun, we must subtract more energy from the ball-sun system. The closer the ball is to the sun, the greater its kinetic energy will be, but it will have *less* potential energy and *less* total energy. In other words, the closer the ball is to the sun, the more bound the system must become. Our ball must lose energy to the environment to be in an orbit closer to the sun and must gain energy from the environment to raise the orbit.

If we extract all the energy we can remove from our ball, it will (if indestructible) be at the surface of the sun. It is now at the lowest energy state (negative maximum) it can have. If we now give it just enough kinetic energy to get it back to our position of zero potential energy and zero force, it will attain *escape velocity*. The escape velocity is the speed a particle must have to become completely unbound from a system.

Escape Velocity from the Earth. The escape velocity from earth is ≈ 11 km/sec. This means that if a spacecraft attains this speed initially

Figure 3–9

at the Earth's surface, it would (neglecting air resistance) escape from the Earth. At any speed less than this, it would return to Earth.* Therefore, to put a spacecraft into orbit with less energy than the escape energy, we inject it into an orbit *above* the Earth's surface, not *at* the Earth's surface.

THE EXPLORATION OF THE SOLAR SYSTEM

Man is in the process of exploring the solar system and may well try to reach the stars. Let's look at some of the problems that will face man in his journey to outer space.

First, we need a way in which to measure the strength of the force of gravity anywhere we happen to be in space. Whenever we get far from the surface of the Earth, many other bodies, such as the sun, the moon, and all the planets, will exert more gravitational force than the Earth. Of course, we could use the universal law of gravitation for each body and add up the results, but this would be quite a mathematical chore. The simplest way is to use the field concept. In a field concept, we assume that a body affects the space around it in some manner. For example, the way the Earth affects the space around it is to cause any other body in the space to fall toward the Earth. The field concept is useful because we can measure the total effect of all bodies contributing to the field by simply finding the value of the force on a given mass. If we are near Earth, we can measure the force on a kilogram of mass and find the value of the acceleration due to gravity, that is,

$$\mathbf{g} = \frac{\mathbf{F}}{M} .$$

At the Earth's surface, a kilogram would have a force of 9.8 newtons exerted on it, so the value of g would be: $g = 9.8$ N/1 kg $= 9.8$ m/sec². The direction of the field is the direction of the force on the mass.

The force of gravity on an object is called the weight of the object.

> Gravitational force = weight.

---

* Many of the lighter molecules, such as hydrogen, attain the escape velocity; consequently, the Earth's atmosphere has very little hydrogen.

If we were to travel 4000 miles above the Earth's surface, we would find that the weight or force acting on the mass would be 2.45 newtons and that $g = 2.45$ m/sec². As we travel farther and farther away from the Earth, we would find that the value of the acceleration of gravity would decrease more and more. Therefore, the value of the acceleration of gravity is a good measure of how the Earth is affecting the space around it. In short, the value of the acceleration of gravity is a measure of the strength of the gravitational field. We can use the value of the acceleration of gravity as a "ruler" to measure the gravitational field anywhere we can calculate the force on a unit mass.

Example 1:   On the moon, a 1 kilogram mass weighs approximately 1.6 newtons. What is the strength of the gravitational field there?

Answer: $\mathbf{g} = \mathbf{F}/M = 1.6$ N/1 kg $= 1.6$ m/sec² in direction of the moon.

Example 2:   The value of the gravitational field on Jupiter is approximately 25 m/sec². How much would a 50 kilogram girl weigh on this planet? (A 50 kilogram girl would weigh 110 pounds on Earth.)

Answer: $W = Mg$

$M = 50$ kg

$g = 25$ m/sec²

$W = (50$ kg$)(25$ m/sec²$) = 1250$ N ≈ 280 pounds.

ENERGY CONSIDERATIONS FOR SPACE TRAVEL

One of the big problems in space exploration is that we must carry all the energy used by rockets in the rockets. From the example on momentum in Chapter 2, we learned that a rocket gets its thrust by throwing part of itself away. This severely restricts the total useable energy that a rocket

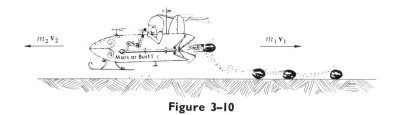

$m_2 v_2$ $m_1 v_1$

Mars or Bust !

Figure 3–10

can carry, since if we increase the mass to throw away (the fuel) we also have to do more work to lift the mass in a gravitational field. Also, the mass used in the final stage of a journey must be carried through all the earlier stages. It would be somewhat analogous to buying a car with a gas tank many times larger than the car. For any given car, there is a limit to how large the gas tank can be and, therefore, a limit to how far the car could travel without refueling. Considering that a rocket must vertically lift all its fuel, one can realize the magnitude of the problem. The problem has been solved somewhat by staging a rocket, but the energy sources for rockets are still limited. Any device (such as a nuclear or ion engine) that will eject mass at a higher velocity will increase the energy capacity.

The overall energy considerations for putting a man on any other planet are:

(1) Energy to escape the Earth's gravitational field.

(2) Energy to compensate for the change in potential energy between an Earth orbit and the chosen orbit.

(3) Energy to descend to the chosen body.

(4) Energy to ascend from the chosen body.

(5) Energy to compensate for the change in potential energy between the chosen orbit and Earth orbit.

(6) Energy to descend from Earth orbit back through the Earth's gravitational field.

Unfortunately, work must be done on the rocket by the rocket engines to accomplish most of these energy requirements.

ENERGY TO ESCAPE THE EARTH'S GRAVITATIONAL FIELD

In order to understand the energy with which a body is bound to Earth, let's look at the way the Earth attracts every body around it. At the surface of the Earth, each kilogram of mass has a force of 9.8 newtons (about 2.2 pounds) exerted upon it by the Earth. The Earth has 9.8 newtons of force exerted upon it by the kilogram of mass, but since the Earth is so large, we can consider it to be not moving (displacement = 0) and the energy gained by the Earth-mass system can be considered as energy gained or lost by the mass. This gravitational force is known as the *weight* of the body, and the acceleration due to the weight is denoted by the symbol g. Newton's law, $F = ma$, can be written using these symbols:

$$W = mg \quad \text{where } W \text{ is the weight in newtons}$$

m is the mass in kilograms

g is the acceleration due to gravity

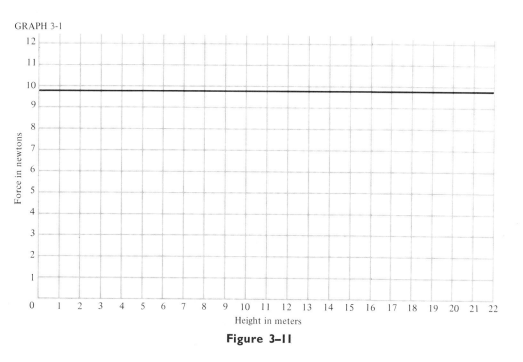

GRAPH 3-1

Force in newtons (vertical axis)

Height in meters (horizontal axis)

Figure 3–11

Do you recall the plot in Chapter 2 of the potential energy? The force near Earth does not vary noticeably with height, and the force versus displacement curve plots a curve like Graph 3–1.

If the force remained constant, as in Graph 3–1, we would never be able to escape from Earth, because the energy necessary to escape from the Earth's gravitational field would be larger than the energy reserves we could put in a rocket or anything else—the work would approach an infinite amount. But, the force does not remain constant; it drops inversely as the square of the distance, according to Newton's law of universal gravitation.

This means that for a body of given weight on the surface of the Earth (1 Earth radius), the weight will diminish to 1/4 of the weight at 2 Earth radii, 1/9 the weight at 3 Earth radii, to 1/16 the weight at 4 Earth radii, and so forth. For example, let's compute the energy necessary to take 1 kilogram of mass from the surface of the Earth to an infinite distance away from the Earth. When we find the energy per kilogram, it is an easy matter to compute the energy for any mass.

Since we are at the Earth's surface, we will declare that to be our position of zero potential energy. We need to know the force as a function of the displacement, and then we need to find the area under the force versus displacement curve to get the energy. Table 3–2 gives the approximate force on the kilogram mass at different displacements.

Now, if we plot the force as a function of the displacement, we get a graph like that in Figure 3–12. Notice that the graph starts at 1 Earth radius, which is the surface of the Earth. The force drops very rapidly at first and levels off as the force approaches zero.

TABLE 3–2 FORCE ON A I KILOGRAM MASS IN THE
VICINITY OF EARTH

DISPLACEMENT FROM CENTER OF THE EARTH		FORCE IN 10^{-1} NEWTONS
6.4×10^6 meters	(surface)	98
12.8×10^6 meters	(2 radii)	24
19.2×10^6 meters	(3 radii)	11
25.6×10^6 meters	(4 radii)	6.1
32.0×10^6 meters	(5 radii)	3.9
38.4×10^6 meters	(6 radii)	2.5
44.8×10^6 meters	(7 radii)	2.0
51.2×10^6 meters	(8 radii)	1.5
57.6×10^6 meters	(9 radii)	1.2
64.4×10^6 meters	(10 radii)	1.0

The energy to take the mass away from the Earth* can be found by
finding the area under the curve. Notice that there is far more area under
$d = 1$ radius to $d = 3$ radii than all other areas combined. After 10 radii,
there is practically no area under the curve. Therefore, if we find the area
for the first 10 radii, we would have a good indication of the energy
necessary to take a 1 kilogram mass that weighs 9.8 newtons (2.2 pounds)
completely away from the Earth.†

GRAPH 3-2

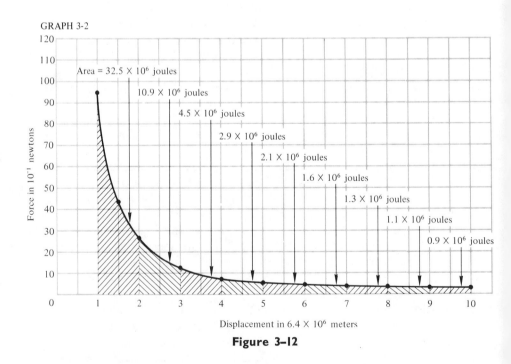

Figure 3–12

* We assume that the Earth does not move.

† Only by using calculus can we find the exact energy. We are limited because we are
unable to draw the curve of an infinite displacement away from the Earth.

To find the area under the curve, the curve was broken up into small rectangles and triangles. The data listed here were obtained.

RADII (1 RADIUS = 6.4 × 10⁶ METERS)	POTENTIAL ENERGY (SURFACE = 0) IN JOULES
1 radius (surface of earth)	0
2 radii	32.5 × 10⁶
3 radii	43.4 × 10⁶
4 radii	47.9 × 10⁶
5 radii	50.8 × 10⁶
6 radii	52.9 × 10⁶
7 radii	54.5 × 10⁶
8 radii	55.8 × 10⁶
9 radii	56.9 × 10⁶
10 radii	57.8 × 10⁶

To put a 1 kilogram mass at a distance of 10 Earth radii would require 57.8×10^6 joules of energy. Although this might seem like a lot of energy, the electric companies sell this much energy for less than 50 cents. The energy to put it to an infinite number of Earth radii would not be much more than this, as Graph 3–3 illustrates.

GRAPH 3-3

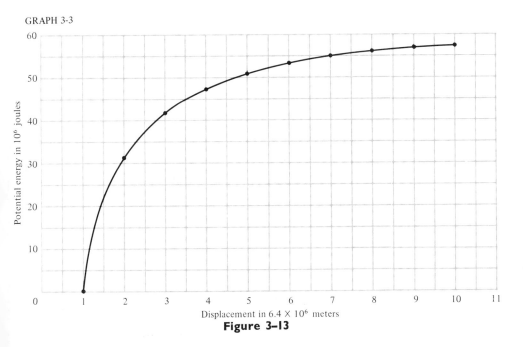

Figure 3–13

The curve flattens out at about 8 radii, and the slope of the energy curve is almost zero, which tells us that it will take almost zero energy to increase the displacement. In fact, the energy to take a 1 kilogram mass

far enough away to completely escape the Earth's gravitational field*
is approximately 62.7×10^6 joules. If we wanted to know the energy to
take any mass *m* away, all we would have to do is to multiply *m* in kilograms by the energy per kilogram. This, of course, is the minimal energy,
because a rocket has to carry its own fuel and oxygen.

Example: How much energy would it require to take a 735 kilo-
gram† spacecraft completely away from the gravitational
field of the Earth?

   Answer: Energy $= (735 \text{ kg})(62.7 \times 10^6 \text{ J/kg})$

   energy $= 46 \times 10^9$ joules.

What happens when we escape the Earth's gravitational field? We
go into orbit around the sun. Since all other planets are also in orbit
around the sun, if we wish to visit a planet in an orbit near us, such as
Venus or Mars, the energy requirement is not large, since we are already
93 million miles from the sun and the force of the sun on a spacecraft
would be small. In fact, the force of the sun on a 735 kilogram spacecraft
is only 4.4 newtons, or approximately one pound. The force of 4.4
newtons would be sufficient to keep our spacecraft in orbit around the
sun at about the same distance the Earth is from the sun.

ENERGY TO CHANGE ORBITS

Mars is farther from the sun than Earth is. To travel to Mars, a
spacecraft must become less bound to the sun, so we must put energy into
what is now the sun-spacecraft system. In order to do this, the spacecraft
must have a speed relative to the sun slightly greater than the orbital speed
of the Earth (3×10^4 meters/sec) and in the same direction. Let's look
at the situation from a time just before lift-off. The spacecraft has a
velocity of zero relative to the Earth, since it is attached to the Earth, but
a velocity of 3×10^4 m/sec relative to the sun (Figure 3–14).

After lift-off, the rocket is given a velocity of 11 km/sec relative to
the Earth. This is enough kinetic energy for the spacecraft to overcome
the binding energy of the Earth and thereby escape the Earth, but all the
kinetic energy has been converted to potential energy. So, the rocket's
velocity relative to Earth is zero, but the orbital speed of the spacecraft
around the sun is 3×10^4 m/sec, the same as the Earth's (Figure 3–15).

Now, if we wish to go to Mars, our spacecraft must be given some
additional speed and launched in the same direction as the Earth moves

---

   * By "escaping" a gravitational field, we mean we go to a point in space where a mass
would not return to the body causing the field.
   † A 735 kilogram spacecraft would weigh 7200 N or 1600 pounds on Earth.

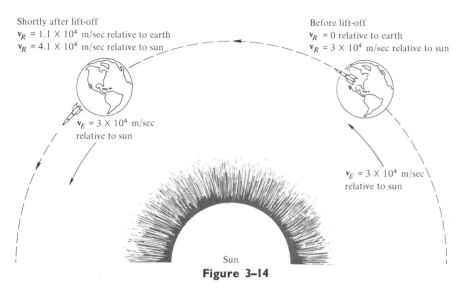

Shortly after lift-off
v_R = 1.1 × 10⁴ m/sec relative to earth
v_R = 4.1 × 10⁴ m/sec relative to sun

Before lift-off
v_R = 0 relative to earth
v_R = 3 × 10⁴ m/sec relative to sun

v_E = 3 × 10⁴ m/sec
relative to sun

v_E = 3 × 10⁴ m/sec
relative to sun

Sun

Figure 3–14

around the sun (counterclockwise to an observer above the plane of the solar system). An additional speed of only 600 m/sec is sufficient to give the spacecraft the necessary energy to approach the orbit of Mars.

If we want to go to Venus, we must subtract energy from the spacecraft. We can do this by giving the spacecraft an additional speed of approximately 600 m/sec, but firing the rocket in the direction opposite the direction of the Earth's orbit (clockwise to an observer above the plane of the solar system). This extracts enough energy to have the spacecraft take the lower energy state to the orbit of Venus.

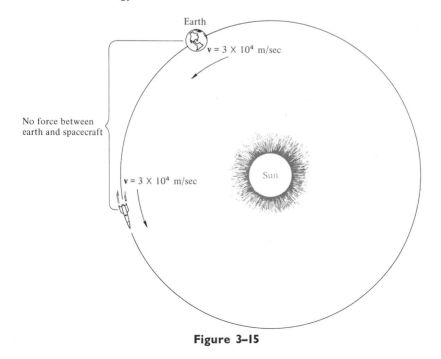

Earth

v = 3 × 10⁴ m/sec

No force between
earth and spacecraft

v = 3 × 10⁴ m/sec

Sun

Figure 3–15

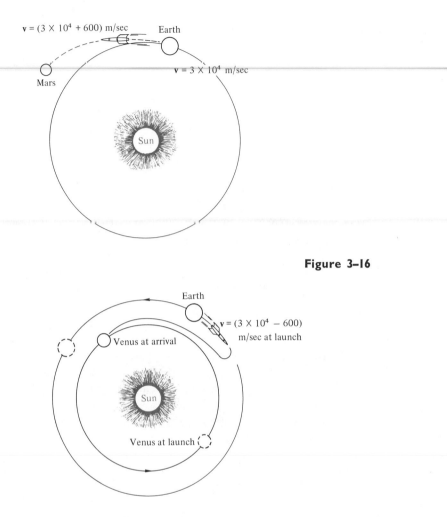

Figure 3–16

We do not have the necessary energy source available in a rocket engine at the present time to escape the solar system. The exploration of regions past the boundaries of the solar system will probably have to wait until nuclear engines for rockets are developed.

ENERGY TO DESCEND TO A PLANET

Whenever the spacecraft gets in the vicinity of another planet, the gravitational attraction of the planet becomes significant, so that the spacecraft increases its speed as it approaches the planet. (The total energy is positive relative to an observer in space.) The kinetic energy of the spacecraft increases as its potential energy relative to the planet decreases, but the total remains constant (and is still positive). Let's use Mars for our example. The spacecraft is aimed to slightly miss Mars, and at some point we must extract energy—that is, make the total energy negative—in order for the spacecraft to be captured by Mars. Otherwise, the craft will be turned in its trajectory but will go into permanent orbit about the sun.

After being captured by Mars's gravitational field, the spacecraft becomes a satellite of the planet and circles it forever if the spacecraft-planet system's energy does not change.

In order to save energy, the main spacecraft would stay in a "parking orbit" above the surface of Mars and an exploration module would descend to the surface. The energy requirement to do this can be computed by finding the binding energy (PE + KE) of the module in the parking orbit (relative to an observer in space) and the binding energy ($PE_1 + KE_1$) on the surface (relative to an observer in space) and subtracting the two energies to find the change in the energy. Another way to compute the energy* is to let the surface of Mars be the point of zero energy and compute the amount of energy necessary to put the module into orbit from the surface of the planet. When we find this energy, we have found the energy requirement to descend, since the energy requirement to descend from the parking orbit and to ascend to the parking orbit are equal.

Since we would like to know many of the energy requirements, let us find the energy necessary to bring one kilogram of mass from the surface of Mars to any displacement up to 10 radii by the same method we used with the Earth. That is, we find the potential energy per kilogram for the first 10 radii. The gravitational field at the surface of Mars is 3.9 m/sec², so the force on 1 kilogram of mass would be 3.9 newtons. Table 3–3 shows the force on a 1 kilogram mass for the first 10 radii.

**TABLE 3–3   FORCE ON A 1 KILOGRAM MASS
IN THE VICINITY OF MARS.**

DISPLACEMENT FROM CENTER OF MARS	FORCE IN 10^{-1} NEWTONS
3.4×10^6 meters 1 radius (surface)	39
6.8×10^6 meters 2 radii	9.8
10.2×10^6 meters 3 radii	4.3
13.6×10^6 meters 4 radii	2.5
17.0×10^6 meters 5 radii	1.6
20.4×10^6 meters 6 radii	1.1
23.8×10^6 meters 7 radii	0.8
27.2×10^6 meters 8 radii	0.6
30.6×10^6 meters 9 radii	0.5
34.0×10^6 meters 10 radii	0.4

Graph 3–4 shows the force versus displacement curve. The area under the curve was found by breaking the area up into small rectangles and triangles and computing the energy to each point of displacement. A potential energy versus displacement curve was then plotted in Graph 3–5.

---

* Without calculus, this is the only way we can compute the energy in this case.

GRAPH 3-4

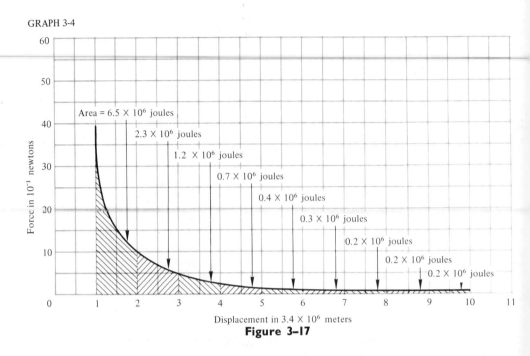

Figure 3–17

We now have the energy per kilogram for the first 10 radii, namely, 12.1 × 10⁶ J/kg. Note that the energy curve flattens out just as it did for the Earth, which tells us that after 10 radii it takes very little energy to remove the kilogram mass completely away from the gravitational field (actually, only 1.1 × 10⁶ joules more). Now imagine that we want a

GRAPH 3-5

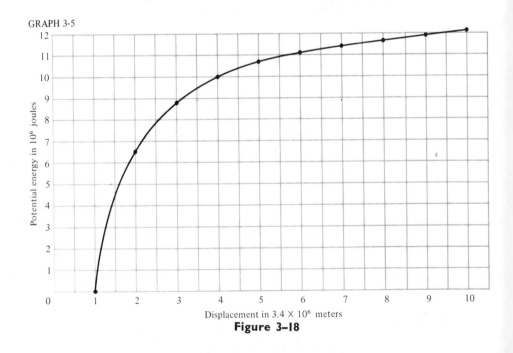

Figure 3–18

spacecraft to take a parking orbit 3 radii from Mars (2 radii above the surface). Our total energy per kilogram in the parking orbit would be:

Total energy = potential energy + kinetic energy.

We can read the potential energy from Graph 3–5 as 8.8 × 10⁶ joules at 3. radii.

In order to find the kinetic energy, we first must find the speed. The speed can be found by the centripetal force formula:

$$F = \frac{mv^2}{R} \quad \text{where } m = 1 \text{ kilogram}$$

$$F = 4.3 \times 10^{-1} \text{ newtons at 3 radii}$$

$$R = 10.2 \times 10^6 \text{ meters}$$

$$v = ?$$

Using these data, a speed of ≈ 2.1 × 10³ m/sec was obtained. The kinetic energy can now be computed, as follows:

$$\text{KE} = \tfrac{1}{2}mv^2 \quad \text{where } m = 1 \text{ kilogram}$$

$$v = 2.1 \times 10^3 \text{ m/sec}$$

$$\text{KE} = \tfrac{1}{2}(1 \text{ kg})(2.1 \times 10^3 \text{ m/sec})^2$$

$$\text{KE} = (\tfrac{1}{2} \text{ kg})(4.4 \times 10^6 \text{ m/sec}) = 2.2 \times 10^6 \text{ joules}$$

The total energy is the potential energy plus the kinetic energy:

Potential energy + kinetic energy

$$8.8 \times 10^6 \text{ J} \quad + \quad 2.2 \times 10^6 \text{ J} \quad = 11 \times 10^6 \text{ J}.$$

The total energy per kilogram at 3 radii above Mars relative to the surface of Mars is 11 × 10⁶ J. The potential energy necessary to escape the gravitational field is 13 × 10⁶. Ignoring the kinetic energy we had when we first entered the gravitational field, we can subtract the two energies just given to find the energy per kilogram that we must subtract from our spacecraft to get it into the parking orbit. Subtracting:

Energy (at infinity) − energy (at 3 radii) = energy extracted

$$13 \times 10^6 \text{ J/kg} \quad - \quad 11 \times 10^6 \text{ J/kg} \quad = 2 \times 10^6 \text{ J/kg}$$

We had to extract 2 × 10⁶ joules of energy for every kilogram of mass in the spacecraft to get it bound to Mars at 3 radii.

It will take far more energy per kilogram to descend from the parking orbit to the surface than it did to descend from infinity to the parking orbit, but less energy is required to send a light exploration module to the surface and back than to land the complete heavy spacecraft.

The energy per kilogram necessary to send the exploration module to the surface is the total energy per kilogram at 3 radii that we have already computed, 11×10^6 joules.* After the exploration, the same amount of energy must be spent getting the module back to the spacecraft. After the module returns from the surface, the journey home begins.

First, the spacecraft must be given enough energy to overcome the binding energy at 3 radii. This will require the same expenditure of energy necessary to get us into the parking orbit, 2×10^6 joules. At 3 radii, the binding energy is not large (note the area under the force versus displacement curve from two radii on Graph 3–4 is small compared to the first 2 radii). After escaping the gravitational field, the spacecraft is still traveling at the orbital speed of Mars around the sun in a counterclockwise direction. It will orbit the sun forever unless it is given an additional speed in a clockwise direction, so that it will fall into a lower orbit. If the right amount of energy is extracted in this manner, the spacecraft will approach Earth's orbit.

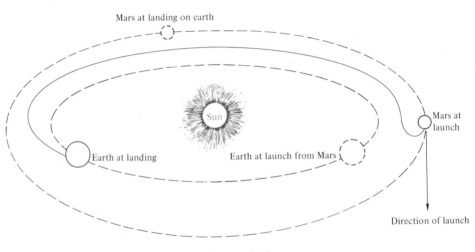

Figure 3–19

After entering Earth's gravitational field, the spacecraft increases its speed toward the Earth (as the potential energy decreases). The spacecraft increases its speed until it encounters the atmosphere, which creates a frictional drag. Frictional forces extract energy from the spacecraft. If no energy is extracted, the spacecraft will bypass the Earth and orbit the sun. If too much energy is extracted over a short period of time, the craft will burn up in the Earth's atmosphere. Therefore, the spacecraft is aimed for the "window" or trajectory that will extract the proper amount of energy without overheating the spacecraft and will bring the spacecraft back to Earth. This has been done many times by the astronauts in the Apollo program.

---

* Mars has some atmosphere, so friction could be used to slow the rocket to some extent.

JOURNEY TO THE STARS

At the present time we do not have the rocket hardware and appropriate fuel to send a spacecraft to the stars, but someday nuclear rocket engines will provide almost unlimited energy reserves. Will man then visit the stars?

In order to take such a long journey, the spacecraft must be a closed ecological system, able to recycle body waste and to produce air, food, and water. In addition, the spacecraft must be able to travel at much faster speeds than are now possible in order to reach even the nearest stars. For example, a spacecraft traveling 22 km/sec, which is fast by present standards, would take 115,000 years for a round trip to Alpha Centauri, the closest star to our solar system.

When man develops a rocket engine with practically unlimited energy sources, a rocket could approach the speed of light (3×10^8 m/sec), and a space traveler could take advantage of time dilation. In time dilation, the clock on a spaceship would run slower than a clock on Earth, so that perhaps a 100 year time span as measured by people on Earth would only be a 4 or 5 year time span to the space travelers* (see Chapter 8).

The magnitude of space is so vast that even with rockets of unlimited energy reserves, man could only "scratch the inner surface" of space around the solar system. The estimated distance to the center of our galaxy (the Milky Way) is 3×10^{20} meters, and it would take 30,000 years traveling at the speed of light to reach the center. But our galaxy is only one of many millions of galaxies. The most distant object man has detected in the universe is estimated to be in the order of 10^{26} meters, and a spaceship traveling at the speed of light would take 10 billion years to travel this distance, so it seems that man is to be limited to a relatively small volume of the universe throughout his existence—but who knows?

EXERCISES

1. What is the gravitational potential energy of a 10 kilogram mass 50 meters above the Earth (if we assume the Earth's surface to be the position of zero potential energy)?

   Figure 3–20 shows a velocity/time curve of a 10 kilogram mass falling on a distant planet.

---

* This would certainly bring about some complex family problems if a space traveler left his wife and children on Earth.

Figure 3–20

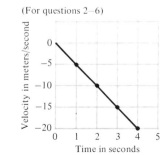

(For questions 2–6)

2. What was the acceleration of gravity on this planet?

3. How much did the object weigh?

4. From what height did the object fall?

5. What was the kinetic energy of the object at $t = 4$ sec?

6. What was the potential energy at $t = 0$?

(For questions 7–10)

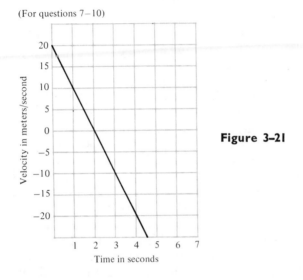

Figure 3–21

Figure 3–21 shows a ball being thrown vertically into the air at the Earth's surface.

7. How long in time did the ball rise?

8. How high in the air did it rise?

9. What was the acceleration of gravity?

10. At what time was the ball back at the starting point?

11. What is the centripetal force on a 25 kilogram mass traveling 10 m/sec in a circle of radius = 5 meters?

(For questions 12–14)

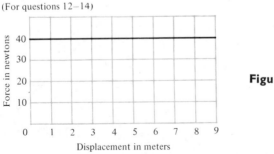

Figure 3–22

Figure 3–22 shows the force/displacement curve of a mass being lifted above the Earth's surface.

12. What was the potential energy at 5 meters?

13. What was the force at 3 meters?

14. Plot a potential energy versus displacement curve for the first 5 meters.

15. What is the acceleration of gravity if 5 kilograms weigh 10 newtons?

16. At a certain point in space, a 5 kilogram mass accelerates 10 meters/sec². What is the weight of the body at this point?

17. A body that weighs 10 newtons accelerates 5 m/sec². What is the mass of the body?

Figure 3–23

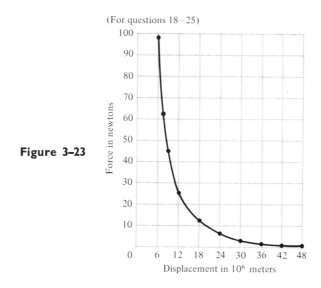

(For questions 18–25)

Force in newtons

Displacement in 10⁶ meters

Figure 3–23 shows the approximate weight of a 10 kilogram mass as a function of displacement from the center of the Earth if we assume 1 radius $= 6 \times 10^6$ meters.

18. How far from the Earth's center would the mass have to be in order to weigh 20 newtons?

19. What is the weight of the object at 6 radii from the center of the Earth?

20. How much energy is required to move the mass from 6×10^6 m to 12×10^6 m?

21. Plot a potential energy curve for the first 4 radii from the center.

22. If a 10 kilogram satellite is in a circular orbit at 5 Earth radii, what is the centripetal force on the satellite?

23. What would be the speed of the satellite at 5 radii?

24. What would be the kinetic energy of the satellite?

25. What is the total energy of the satellite?

26. How much would a 100 kilogram (220 pound) football player weigh on the surface of Mars? (See Table 3–1.)

27. How much energy would it take to send a 1000 kilogram satellite completely away from Earth?

28. With what speed would a 1000 kilogram satellite enter the Earth's atmosphere if it fell from an infinite distance from the Earth? (Assume atmosphere is at Earth's surface.)

29. How far do you think man will travel from Earth in the next 50 years?

30. Why do you think Mars has a far less dense atmosphere than Earth? Why does the moon not have any atmosphere?

ELECTRICITY AND MAGNETISM

ELECTRIC CHARGES AT REST AND IN MOTION

ELECTROSTATICS

In the wintertime, have you ever scooted across a car seat and received an unpleasant shock as you touched the door handle? Have you ever combed your hair in darkness on a cold night and seen sparks? Have you pulled clothes from a dryer and found some of them sticking together? Have you rubbed a fluorescent tube with plastic food wrap in the dark and watched the tube glow? All of these illustrate static electricity. Static electricity deals with electric charges at rest.

If you rub two hard rubber rods with cat fur (not containing the cat), the rods will repel each other. Two glass rods that have been rubbed with silk will also repel each other. If the experiment is done on a cold, dry day, other facts will be observed:

(1) The glass rod will attract the rubber rod.
(2) The silk will repel the rubber rod.
(3) The cat fur will repel the glass rod.
(4) The cat fur will attract the silk.

From these facts, you can deduce that there must be two kinds of electricity. Since two identical objects rubbed with the same material must have the same kind of charge, we can make the statement, "Like charges repel each other and unlike charges attract each other." The two types of electricity we denote as positive and negative. The first American physicist, Benjamin Franklin, defined the charges as follows: (1) the kind of electricity that is on a hard rubber rod after the rod has been rubbed with cat fur is *negative electricity* (2) the kind of electricity that is on a glass rod after the rod has been rubbed with silk is *positive electricity*.

101

It is suspected, although unproven, that the total electric charge in the universe is zero; in other words, for all the negative charges there are an equal number of positive charges. When we rub a neutrally charged hard rubber rod with cat fur, which is neutrally charged, we simply separate some positive charges from the negative charges. Any device that separates charges is called a generator of electricity.

For many years it was thought that electricity flowed, like a fluid, onto or away from an object, making it have a positive or negative charge. In modern times, several experiments have proven that electricity is particulate or "grainy" in nature. There is a particle that has a smallest charge, called the electron, which has a certain mass and a certain charge.

An electron has the smallest possible negative charge.

An electron has a mass of 9.11 × 10⁻³¹ kilograms.

Had it been known earlier that the electron was the smallest unit of charge, the electron might have been the fundamental unit of charge. However, the system of electrical measurement was well established when the electron was discovered, so the fundamental unit of electrical charge is a much bigger unit, called the coulomb.

A coulomb is 6.24 × 10¹⁸ electrons.

Although a coulomb is a convenient unit to use in current electricity, it is a fantastically enormous unit to use in electrostatics. The force between two charges of 1 coulomb 1 meter apart would be 9 billion newtons (2 billion pounds). The charges on even the most highly charged objects are equal to only a small fraction of a coulomb. Therefore, in electrostatics we use the prefixes *micro*, *pico*, and *nano*, and so forth to give the unit of charge. Since the coulomb is equal to 6.24 times 10¹⁸ electrons, the quantity of charge of one electron, is

$$\frac{1 \text{ coulomb}}{6.24 \times 10^{18} \text{ electrons}} = -1.60 \times 10^{-19} \text{ coulomb.}$$

This figure is the smallest unit of charge.

Everything that has a charge would have some integral (whole number) multiple of the smallest charge. That is, nothing could have a charge of 1½ or 5¼ electrons. Whenever something is made up of individual packets like the electric charge of the electron, we say that it is *quantized*.

The proton is a particle that has approximately 1840 times the mass of the electron and has an equal but opposite charge. That is:

The charge of one proton = +1.60 × 10⁻¹⁹ coulomb.

Example 1:   If you had 6.24 billion electrons, how many coulombs would you have?

Answer: $(6.24 \times 10^9 \text{ electrons})/6.24 \times 10^{18} \text{ electrons}/C = 1 \times 10^{-9}$ C.

Example 2:   How many electrons are there in a nanocoulomb?

Answer: $(6.24 \times 10^{18} \, e/C)(10^{-9} \, C) = 6.24 \times 10^9$ electrons.

THE INVERSE SQUARE LAW

Suppose that we take two rubber rods, *A* and *B*, and keep them exactly 1 meter apart. We next put a charge of 10 microcoulombs on rod *A* and a charge of 10 microcoulombs on rod *B*, and then measure the force between them.

We repeat the experiment, changing the charge on both rod *A* and rod *B*, and tabulate the results (Table 4–1).

TABLE 4–I

CHARGE ROD A (in 10^{-5} coulomb)	CHARGE ROD B (in 10^{-5} coulomb)	FORCE NEWTONS (all values × 0.9)
1	1	1
2	1	2
2	2	4
3	1	3
3	2	6
3	3	9
4	1	4
4	2	8
4	3	12
4	4	16

Table 4–1 shows that the product of the two charges is proportional to the force. Therefore, we can write: $F \propto Q_1 Q_2$.

$F \propto Q_1 Q_2$

Suppose that we repeat the experiment, but this time we keep the charge on rod *A* and rod *B* constant and change the distance, then tabulate the results as in Table 4–2. Let the charges on each rod be 10 microcoulombs.

TABLE 4–2

DISTANCE IN METERS BETWEEN ROD A AND ROD B	FORCE IN NEWTONS (all values × 0.9)
1	1
½	4
⅓	9
¼	16
⅕	25

If we graph the results, we get a curve like the inverse square law for the force of gravity; therefore, from the graph we can deduce that the force is inversely proportional to the square of the distance between the rods, and we can write $F \propto 1/R^2$.

$F \propto 1/r^2$

GRAPH 4-1

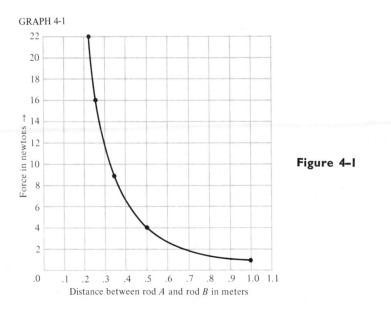

Figure 4–1

Since the force is directly proportional to the charges, we can combine both statements into one proportionality, namely, $F \propto Q_1 Q_2 / R^2$.

This equation implies that $F = k(Q_1 Q_2 / R^2)$, where k is the connecting constant, however, we must find the value of the connecting constant. To do this, we simply do the experiment again, knowing all values except the constant k. If we do this using any of the foregoing data we will find that the constant k is equal to 9.0×10^9 newton meter²/coulomb². Therefore,

$$F = 9 \times 10^9 \left[\frac{\text{newton meter}^2}{\text{coulomb}^2} \right] \frac{Q_1 Q_2}{R^2}$$

where Q_1 = one charge in coulombs

Q_2 = other charge in coulombs

R = distance between the charges in meters

This relation is the fundamental force equation for electrostatics or charges at rest. It is called Coulomb's Law.*

* It is recommended that Experiment 5 in the lab manual be done at this time.

The preceding equation gives us only the magnitude of the force. The direction of the force is parallel to a line joining Q_1 and Q_2 and is attractive if Q_1 and Q_2 are unlike charges and repulsive if they are like charges.

The coulomb force is much stronger than the gravitational force. The electrostatic force between an electron and a proton in a hydrogen atom is about 10^{39} times stronger than the gravitational force.

Example: What is the magnitude of the repulsive force of two protons in a nucleus if the charge on each proton is 1.6×10^{-19} coulomb and the protons are 3×10^{-15} meters apart?

Answer:

$$F = 9 \times 10^9 \left(\frac{Nm^2}{C^2}\right) \frac{Q_1 Q_2}{R^2} \quad \text{where } Q_1 = 1.6 \times 10^{-19}\ C$$

$$\text{or } 16 \times 10^{-20}\ C$$

$$Q_2 = 1.6 \times 10^{-19}\ C$$

$$\text{or } 16 \times 10^{-20}\ C$$

$$R = 3 \times 10^{-15}\ m$$

$$F = (9 \times 10^9)\frac{Nm^2}{C^2}\frac{(16 \times 10^{-20}\ C)^2}{(4 \times 10^{-15})^2} = 14.4 \text{ newtons.}$$

THE ELECTRIC FIELD

We used the value of the acceleration of gravity, $\mathbf{g} = \mathbf{F}/m$, to measure the effect of the gravitational field. We wish to have a similar measure for an electric field. We assume that a charge in space will change the space around it in some fashion. We say that the charge sets up an electric field, and to measure the strength of this field, we define the electric field intensity as equal to the force per unit positive charge. That is:

$$\mathbf{E} = \frac{\mathbf{F}}{+Q} \quad \text{where } \mathbf{E} \text{ is electric field intensity in newtons/coulomb}$$

$$\mathbf{F} \text{ is force in newtons}$$

$$Q \text{ is electric charge in coulombs}$$

The electric field $\mathbf{E}$ is a vector, since force is a vector and Q is a scalar. The direction of the $\mathbf{E}$ field at any point is the direction of the force on a positive charge at the point. The advantage of having the concept of an $\mathbf{E}$ field is that we can measure the strength of the electric force without knowing anything about the charges that cause the force. Since forces due to many charges can be quite mathematically complex, it is usually easier to use the field concept.

In an electric field, we use *imaginary lines* to portray the direction and strength of the field. These lines begin on positive charges and end on negative charges. The direction of the line is the direction of the force on a positive charge. The closer the lines, the stronger the field. If lines are like those in Figure 4–2, the field weakens from *a* to *b* and a positive test charge would have more force on it at *a* than at *b*. If the lines are like those in Figure 4–3, the field strengthens from *a* to *b* and a positive test charge would have more force on it at *b* than at *a*. If the field lines are like those in Figure 4–4, the field is constant and the force on a positive charge would be constant everywhere in the field.

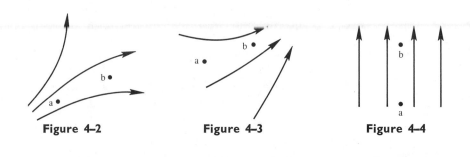

Figure 4–2   Figure 4–3   Figure 4–4

Example: A charge of 5×10^{-6} coulombs experiences a force of 10 newtons at a certain spot in space (point A). (a) What is the electric field at point *A*? (b) Is the force on a proton greater at *A* or at *B*?

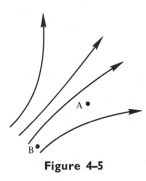

Figure 4–5

Answer: (a) $\mathbf{E} = \mathbf{F}/Q = 10$ newtons/5×10^{-6} coulomb $= 2 \times 10^6$ newtons/coulomb. (b) Greater at *B*.

POTENTIAL DIFFERENCE

One of the reasons for our interest in electricity is that it will transmit energy for us. The electric field is a conservative force field; therefore, when we work on a charged particle against an electric field, all the work that we do on the charged particle can be regained by the particle doing work for us. Work is done by some outside agent on a charged particle at an energy source and is obtained from a charged particle at an energy sink. In order to compute the amount of work given to or taken from a charged particle, we need a unit of potential energy difference, and we define this unit in the following manner: The difference in potential between two points *A* and *B* is the work done on a unit positive charge moving it from *A* to *B*. That is:

$$V = \frac{W}{+Q}$$ where *V* is potential difference in volts

W is work done in joules

Q is charge in coulombs

A positive potential difference between *B* and *A* means that the charge has a greater amount of potential energy at position *B* than it did at position *A*.* We say *B* is at a higher potential than *A*. Some outside agent had to do work on the charge against the forces of the electric field to get the charge from position *A* to position *B*. For example, if the agent

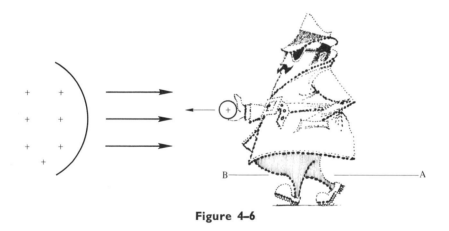

Figure 4–6

did 5 joules of work on 1 millicoulomb (10^{-3} C) in taking it from *A* to *B*, the potential difference between *B* and *A* is:

$$V = \frac{W}{Q} = \frac{5 \text{ joules}}{10^{-3} \text{ coulomb}} = 5000 \text{ J/C or } +5000 \text{ volts.}$$

A negative potential difference means that the charge has less potential energy at position *B* than it did at *A*. In this case, the force of the electric field works on the charge and the charge either accelerates or loses its energy to some outside agent. For example, if a 1 millicoulomb charge

---

* Actually, we are talking about a test charge-electric field as a system. As in the case of the gravitational field, we assume that only our test charge moves.

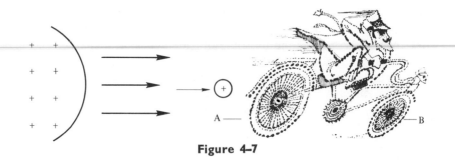

Figure 4-7

did 5 joules of work on some outside agent from *A* to *B*, then the difference of potential between point *B* and point *A* is:

$$V = \frac{W}{Q} = \frac{-5 \text{ joules}}{10^{-3} \text{ coulomb}} = -5000 \text{ volts.}$$

We increase the electric potential of a charge by moving it against the electric force of the field; we decrease the potential by letting the charge move in the same direction as the electric force.

Example 1: Ten joules of work is performed on a $+5 \times 10^{-3}$ coulomb charge in moving the charge against the electric field from point *A* to point *B*. What is the potential difference between the two points, and which point is at the higher potential?

Answer: Work is done on the charge, therefore, point *B* will be at a higher potential than point *A*.

$$\text{Increase in potential} = \frac{\text{work}}{\text{charge}} = \frac{10 \text{ joules}}{5 \times 10^{-3} \text{ coulomb}}$$
$$= 2000 \text{ volts.}$$

Example 2: For every 50 coulombs of charge that flow through an electric motor, the motor performs 10,000 joules of work. What is the potential drop across the electric terminals of the motor? (Assume that the motor is 100 per cent efficient.)

Answer: In this case, the work is done by the electric charges and the electric potential will decrease.

$$\text{Increase in potential} = \frac{-10,000 \text{ joules}}{50 \text{ coulombs}} = -200 \text{ volts.}$$

A negative potential difference is called a voltage drop.

Example 3: What will be the kinetic energy increase of an electron if it has been accelerated through a potential difference of 20 million volts? (Assume that $e = 1.6 \times 10^{-19}$ coulomb.)

Answer: In this case, the charge is released and the electric field works on the electron, increasing its kinetic energy. The increase in kinetic energy is equal to the decrease in the potential energy of the field electron system.

$$\text{Potential} = \frac{\text{work}}{\text{charge}} \quad or \quad \text{work} = (\text{potential})(\text{charge})$$
$$\text{(in volts)}$$

$$\text{work} = (20 \times 10^{6})(1.6 \times 10^{-19}) \frac{\text{(joules)}}{\text{(coulombs)}} \text{(coulombs)}$$

$$\text{work} = 32 \times 10^{-13} \text{ joules lost by the system, therefore the electron gained this much kinetic energy.}$$

CAPACITANCE

A capacitor (or condensor) is a device that stores electric charges. It is a most useful electrostatic device, since a considerable number of capacitors are in every electronic device—radios, television sets, record players, computers, and so on. Electric charges never flow through a capacitor, but can flow in the wires leading up to and away from the capacitor. The simplest capacitor (called a parallel plate capacitor) is two parallel conductors separated by an insulator, as shown in Figure 4–8. When charges flow toward one plate of a capacitor, like charges flow away from the other plate because of the electrostatic repulsion.

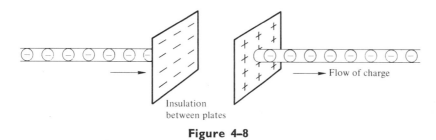

Insulation
between plates

Figure 4–8

Since charges are being separated, there is a potential difference between the plates, which is opposite to the potential difference that is causing the electrons to flow on the plate. One plate becomes more and more negative (repelling other negative charges trying to flow onto the plate) and the other plate becomes more and more positive (attracting electrons trying to flow from the plate). Eventually, the forces pushing the electrons on the plate are just equal to the forces repelling the electrons on the same plate, and the flow of electrons ceases. The potential difference across the plates of the capacitor is now equal but opposite to the potential difference that was causing the flow. The capacitor is now fully charged for the particular voltage. Since one plate has just as many negative charges as the other has positive charges, and electric field lines start on positive charges and end on negative charges, the electric field will be constant between the plates and will be zero outside. If the voltage between the capacitor plates reaches a sufficiently high value, the insulation between the plates breaks down and a spark jumps from one plate to the other, discharging the capacitor. Lightning is an example of a capacitor discharge—the clouds act as one plate and the ground (or another cloud) is the other plate.

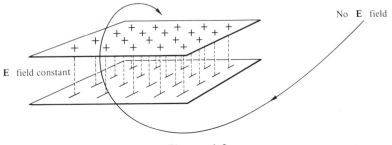

Figure 4-9

Capacitors come in different electrical "sizes." The size of a capacitor is measured by a quantity called *capacitance*, which is defined as the ratio of the charge on one plate to the potential difference across the plates. The unit of capacitance is a farad, but the practical unit is microfarads, since it would take a capacitor the size of a building to be a 1 farad capacitor.

A capacitor is a handy electrical device. It has the ability to discharge small amounts of energy quickly (as in an electronic flash unit) and the ability to accelerate charged particles, and it is useful as a device to discriminate between electrical signals of different frequencies (as in a radio). A parallel plate capacitor is especially useful because a constant electric field of any desired strength can be created between the plates, and the electric field outside the plates is equal to zero. If we had a like device for the gravitational field, we could make objects weigh any amount, make them float, or make them have negative weight, so that they would fall up instead of down. Since the force between masses is attractive, such a device would not be possible.

To create a constant electric field of any desired strength, we use the relation:

$$E = \frac{V}{d}$$ where E is the magnitude of the electric field in volts/meter or newtons/coulomb

V is the potential difference between the plates

d is the distance between the plates

Note that the electric field can also be expressed as volts/meter using the foregoing relation.

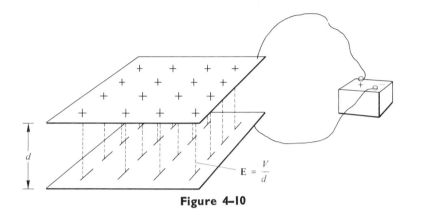

Figure 4–10

Example: What must be the potential difference between the plates of a parallel plate capacitor if a constant electric field of 5000 N/C is needed and the plates are 10^{-2} meters apart?

Answer: $E = 5000 \ V/m$

$V = ?$

$d = 10^{-2} \ m$

$5000 \ V/m = \dfrac{V}{10^{-2} \ m}$

$V = 50$ volts.

In many electronic devices, such as television picture tubes, it is necessary to accelerate electrons to very high velocities and to deflect electrons up and down and left and right by passing them between the plates of a capacitor. Since the force on the electrons is proportional to the electric field, we can give an electron any desired speed or deflect it by any amount by controlling the electric field strength and placement of the capacitor.*

To linearly accelerate an electron, the electron travels from the negative plate toward the positive plate (Figure 4–11). The potential

---

* Electrons can also be deflected magnetically.

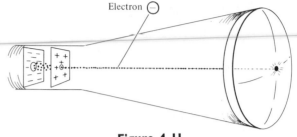

Figure 4-11

energy loss is equal to $(V)(e)$ where V is the potential difference between the plates and e is the charge of an electron. The potential energy loss is the kinetic energy gain of the electron, that is:

$$\text{loss in PE} = \text{gain in KE}$$
$$Ve = \tfrac{1}{2}mv^2$$

Upon arriving at the positive plate, the electron goes through a hole in

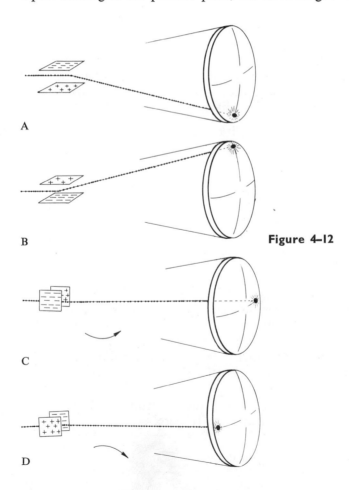

A

B

Figure 4-12

C

D

the positive plate and continues with constant velocity (in the absence of retarding forces).

To deflect an electron downward, the electron goes between the capacitor plates, as shown in Figure 4–12A. The electric field exerts a constant downward force on the electron while it is between the plates, changing the direction of its trajectory. To deflect it upward, the voltage between the plates is reversed, as shown in Figure 4–12B. Figure 4–12C and D show how an electron is deflected to the left or right.

Since an electron can be deflected up and down or left and right, we can guide an electron to any spot we want it to go by changing the voltage on the vertical and horizontal capacitor plates.

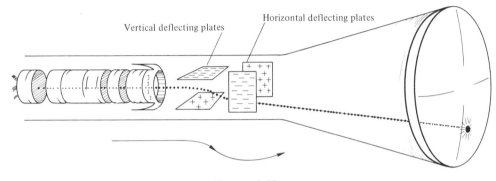

. **Figure 4–13**

Figure 4–13 shows how the elements are placed in the cathode ray tube. Upon striking the screen, electrons cause the phosphor on the screen to glow, thus producing a bright spot. The cathode ray tube is an essential part of some television picture tubes and of the cathode ray oscilloscope, an important research tool with many practical applications. It can be found in every research laboratory, hospital, and television repair shop, and is also used as a stage prop for "monster movies." In a cathode ray (electron stream) oscilloscope, the dot "sweeps" at a constant rate across the screen from left to right. Upon arriving at the right side, the dot is quickly put back to the left, and it repeats the sweep over and over in a regular periodic fashion. This is accomplished by putting a "sawtooth" voltage on the horizontal plates.

For example, suppose we want to sweep the screen in 2 seconds. We will assume that a potential difference of +10 volts between the capacitor plates (called horizontal deflector plates) is sufficient to make the dot hit the screen on the extreme left (Figure 4–14). We decrease the voltage until at $t = 1$ second the voltage is zero and the dot is in the middle of the screen. The voltage continues to decrease (plates are now oppositely charged) until at $t = 2$ seconds the voltage is −10 volts and the dot is on the extreme right. Now we reverse the voltage very quickly, so that the dot will immediately come back over to the left for the next sweep. The time of the sweep can be adjusted to any time. The sweep is usually in terms of

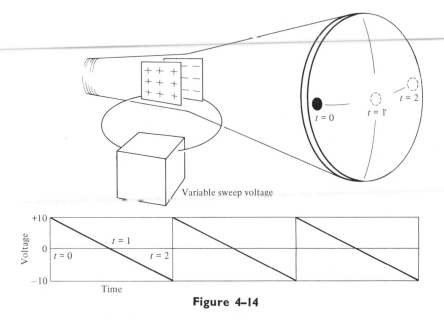

Figure 4-14

fractions of a second, and the term "sweep frequency" is used to indicate the number of sweeps each second.

As the dot sweeps from left to right, a "signal voltage" is put on the vertical deflecting capacitor plates. Many signals can be converted into a voltage that is an exact replica of the signal, whether the signal be a beating heart, a musical tone, or the counting of particles. If the signal repeats in a regular (periodic) fashion, we can adjust the sweep frequency to the frequency of the signal, and a "snapshot" of the signal will appear on the oscilloscope screen. The ability of the oscilloscope to stop and analyze signals makes it a valuable research tool.

In a television picture tube, the intensity of the spot is varied while the spot moves very rapidly back and forth across the screen, painting a complete still picture on the phosphor-coated screen every 1/30 second. With 30 still sequential pictures being flashed on the screen every second, the eye interprets the picture as smooth movements. The voltages on the vertical and horizontal capacitor plates determine precisely where the spot will be at any instant. The voltages on these plates are controlled by the transmitting station. The transmitting station sends the same synchronizing signals to the television camera and to the receiver. In this manner, a piece of picture information on the camera screen is put precisely at the same spot on the receiving screen.

CURRENT ELECTRICITY

An electric current will efficiently transmit energy from one point to another point. In order to make it do this, we work on electric charges at one point and extract work from the charges at another point. The device that works on the charges is called a power source and the device that extracts work from charges is called a power sink or output. Examples of power sources are electric generators, flashlight and car batteries (cells), and solar batteries (cells). Examples of power sinks are light bulbs, electric appliances, and electric motors.

Figure 4–15 illustrates a simplified electric circuit. Note that the charges are already in the conducting medium, which is usually copper or aluminum wire. These materials have electrons that are so loosely bound to the parent atoms that the electrons are called free or conduction electrons.

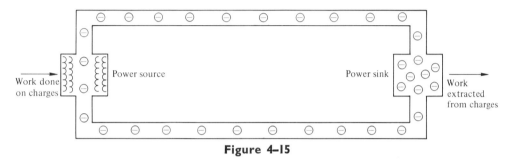

Figure 4–15

The total work that we can extract from the electrons will be limited only by the total work that we do on the electrons. Sometimes we have several devices that extract work from the same charge, as Figure 4–16 illustrates. In this case we say that the devices are connected in series with the power source. It can easily be seen that the charges must go through every device and that the energy of the power source must be divided between the devices. Also, if for some reason one of the devices breaks down or is turned off, the charges stop flowing to all other devices. Series connections are used in many different ways. The most familiar example is that type of Christmas tree light set in which all bulbs go off if one burns out.

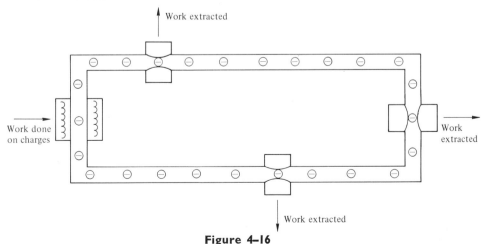

Figure 4–16

Another way the power sources can be connected to the power sinks is through a parallel connection (Figure 4–17). The charges that go through one device will not go through the other device. Each device will extract all the energy of the charges that pass through it. The electrical circuits in a house are in a parallel connection. As you turn on more appliances, there is a corresponding increase in the total electrical current into the house circuits. A fuse or a circuit breaker put in series limits the total current through any particular circuit. The main fuses limit the total current from the transmission line into the house. The house is thereby protected from overloading circuits.

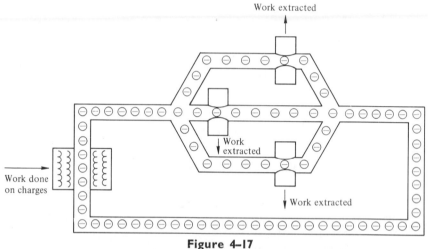

Figure 4–17

In order to compute the amount of work done by the charges, we need to know (1) the number of charges and (2) the energy given up by each charge.

In order to compute the number of charges, we define a unit of current intensity called the ampere:

$$I = \frac{\Delta Q}{\Delta t}$$

where I is the intensity of the current in amperes
ΔQ is the amount of charge in coulombs
Δt is the time in seconds it takes for the charge to pass the point

If we know the current intensity, and the time, the amount of charge passing through a point in the circuit can be computed.

Imagine that you are able to detect the charges going by in an electric circuit (Figure 4–18). Every time you count 6.24×10^{18} electrons, one coulomb has gone by. If 5 coulombs go by in 1 second, the current is 5 amperes. If 5 coulombs go by in $\frac{1}{2}$ second, the current is 10 amperes. Since the forces between the charged particles are so strong, charges will not be able to pile up anywhere in the circuit. The current intensity will be

Figure 4–18

the same everywhere in the circuit unless an alternate or parallel path is provided. If a parallel path is provided (Figure 4–19), the sum of the currents in the separate branches must add up to the current toward or away from the parallel branches.

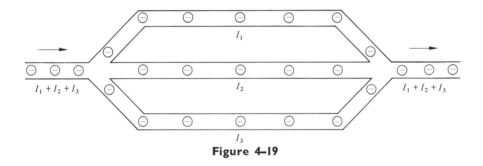

Figure 4–19

Example 1: Fifty coulombs of charge flow past a point in 10 seconds. What is the intensity of the current in amperes?

$$Answer: I = \frac{\Delta Q}{\Delta t} \quad \Delta Q = 50 \text{ coulombs} \quad \Delta t = 10 \text{ seconds}$$

$$I = \frac{50 \text{ C}}{10 \text{ sec}} = 5 \text{ amperes.}$$

Example 2: Graph 4–2 shows the charge that has passed through an

Figure 4–20

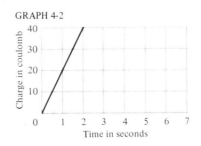

GRAPH 4-2

electrical device as a function of the time. What is the intensity of the current?

$$Answer: \quad I = \frac{\Delta Q}{\Delta t} = \text{slope of line}$$

$$I = \frac{40 \text{ C}}{2 \text{ sec}} = 20 \text{ amperes.}$$

RESISTANCE IN AN ELECTRICAL CIRCUIT

Refer to Figure 4–16 and look a little closer at the agent that extracts energy from the circuit. Actually, the charges lose energy in every part of the circuit. A very sensitive voltmeter could detect a potential difference between every two parts of the circuit; however, the circuit is designed so that most of the potential difference is across the device from which we want to extract work, since we want the device to extract as much work as possible and the connecting wires to extract as small amount of work as possible. The amount of charge that flows through the device is determined by its impedance to the flow of charge. The smaller the impedance, the greater the current for a given voltage.

It is an easy matter to measure the current through a device with an ammeter and to measure the voltage drop or potential difference across the device with a voltmeter. The ratio of voltage to the current is defined as the impedance of the device.

If the device changes the electrical energy fed into it to some form of *heat* energy, the impedance is called a *resistance* to the flow of current, and we state:

$$\text{Resistance} = \frac{\text{potential difference}}{\text{intensity of current}} \quad \text{or} \quad \boxed{R = \frac{V}{I}}$$

where R = resistance in ohms

$\qquad V$ = potential drop in volts

$\qquad I$ = current in amperes

Since most devices are designed to operate with a definite voltage and current, the resistance of a device is usually considered to be constant. The test of whether the resistance is truly constant can be found by plotting a voltage versus current curve.

In Figure 4–21, the curve in Graph A is a straight line, which tells us that the resistance is constant. This very important result is known as *Ohm's Law*. Ohm's Law holds true as long as the voltage versus current curve is a constant. Most metal conductors and commercial resistors obey Ohm's Law within the limits in which they are designed to operate. Graph

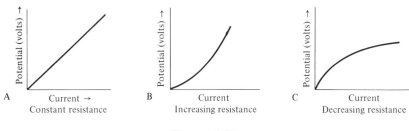

Figure 4-21

B and Graph C show curves in which resistance is not constant, so Ohm's Law does not hold. Many useful electronic devices, such as radio (vacuum) tubes and transistors, work because they *do not* obey Ohm's Law.

Several things determine the resistance of a resistor. Consider three pieces of wire that are identical in length, width, and thickness, but made of different materials: copper, aluminum, and iron. We connect each wire to identical 10 volt batteries and measure the current through each wire. The current through the copper wire is 5.9 amperes, the current through the aluminum wire is 3.8 amperes, and that through the iron wire is 1.0 ampere. When we solve for the resistance, we find:

$$R \text{ (copper)} \qquad 1.7 \text{ ohms}$$

$$R \text{ (aluminum)} \qquad 2.6 \text{ ohms}$$

$$R \text{ (iron)} \qquad 10 \text{ ohms}$$

The wires are of identical proportions, so we conclude that the different values of resistance must be due to the different molecular structure of the different materials.

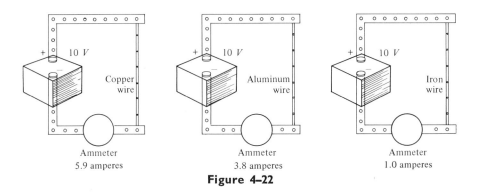

Figure 4-22

If wires made up from many different materials are used in the preceding experiment, we find that the resistance varies over a very wide range. For example, the resistance of fused quartz is about 10^{25} times the resistance of copper. Those materials that have a relatively low resistance, such as metals, are called *conductors*. Those materials that have a very very high resistance, such as quartz, rubber, plastic, and glass, are called *insulators*, and all materials in between, such as silicon and germanium, are called *semi-conductors*.

We could do similar experiments with any one of the materials and show that if we double the length, we double the resistance; if we double the cross-sectional area, we halve the resistance. Therefore, resistance depends upon the kind of material, the length of the wire, and the cross-sectional area of the wire. We can use this information to design a resistor with resistance of any particular value.*

The symbol for a resistor in a circuit is —ᴡᴡ—. The symbol for a direct current power source (battery) is —|‖⊥. The Greek letter Ω is used to abbreviate the word "ohm" and the value of the resistor is usually given on the resistor by a color code or by printing on the resistor.

Example: What is the current through a 5 ohm resistor connected to a 20 volt battery?

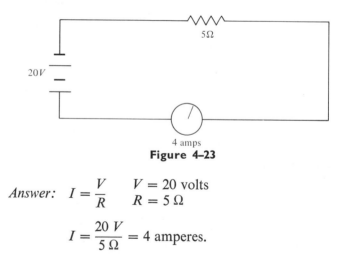

Figure 4–23

$$Answer: I = \frac{V}{R}   \quad V = 20 \text{ volts} \\ \quad R = 5 \text{ } \Omega$$

$$I = \frac{20 \ V}{5 \ \Omega} = 4 \text{ amperes.}$$

ELECTRIC POWER

We cannot count the number of charges going through an electrical device. We use an instrument called an ammeter that will read the current through the device when placed in series with it.

With a voltmeter, we can read the potential difference or voltage drop across the device. Knowing the voltage drop across the device and the

---

* It is recommended that Experiment 6 in the lab manual be done at this time.

current in amperes through the device enables us to find the power used in the device. Remember the definition of power:

$$\text{Power} = \frac{\text{work}}{\text{time}} = \frac{\Delta W}{\Delta t}$$

By a little algebraic manipulation, we can show that the power used by a device is the product of the current through the device and the potential difference across the terminals of the device. That is:

> $P = VI$ where P is the power in watts
>
> V is the voltage drop across the device
>
> I is the current through the device

Sometimes we wish to know the value of the power, and we know only the current and the resistance. We can use Ohm's Law to derive an expression using only these parameters.

$P = VI$, but from Ohm's Law, $V = IR$. Therefore,

$P = (IR)(I) = I^2R.$

Electrical power can be transmitted, with any value of voltage and any value of current, since power $(P) = VI$. Transmission line power losses are equal to $P = I^2R$, where I is the current and R is the resistance of the transmission line. In order to avoid large energy losses that would only heat the transmission lines, electric power companies use very large voltages to transmit power from the generating plant to the cities. For example, to transmit a given amount of power, we can increase the voltage 100 times and reduce the current to 1/100 its original value. The line losses will then be only 1/10,000 as much.

Electrical devices used in houses are designed to work on a particular potential difference (in the United States, the potential difference is usually 110 or 220 volts, depending upon the power requirements). The device is designed so that just enough current will flow through the device to meet the power requirements of the device. One of the power lines is grounded to the earth and is declared to be zero potential. Therefore, a voltage of 110 volts is understood to be a potential difference of 110 volts.

Example 1:   What current must flow through a 55 watt bulb connected to a 110 volt source?

> *Answer:* $P = VI$ $P = 55$ W $V = 110$
>
> $55 = 110\,I$
>
> $\frac{1}{2}$ ampere $= I.$

Example 2:   What is the minimal current that a motor can "draw" if it is connected to a 110 volt line and it must do work at the rate of 770 joules/sec (assuming the motor is 100 per cent efficient)?

Answer: 770 joules/sec = 770 watts; therefore,

$$P = VI$$

$$770 = 110\ I$$

$$7 \text{ amperes} = I.$$

EXERCISES

1. How many electrons flow by in 1 second if there is a current of 1 ampere?

2. What is the electric field intensity if a charge of 1 millicoulomb experiences a force of one newton?

3. What is the increase in potential when 5 joules of work is performed on a 5×10^{-4} coulomb charge?

4. An electric motor is used on a 120 volt line. How much work is done by the charges when 80 coulombs of charge flow through the motor?

5. A proton is accelerated through a potential of 1 billion volts. How much kinetic energy would the proton have? (The charge on a proton is the same as the charge on an electron.)

6. Why is an electrical storm rare in the winter?

7. Work is done on a 1 millicoulomb positive charge where the force on the charge is a function of displacement from a point, as given in Figure 4–24. (a) What work is done on the charge from 0 to 3 meters? (b) What is the increase in potential from 0 to 3 meters? (c) What is the difference in potential between 3 meters and 10 meters?

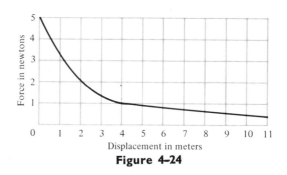

Figure 4–24

8. Figure 4–25 plots the power of an electric motor against time. Calculate the energy dissipated during the first 50 seconds.

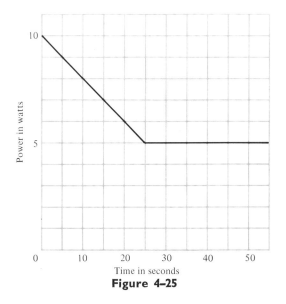

Figure 4–25

9. A resistor in a television set has 2 amperes flowing through it when the voltage drop across the resistor is 200 volts. (a) What is the resistance? (b) What is the power dissipated by the resistor?

10. What is the resistance of a 220 watt bulb designed to operate on a 110 volt line?

11. An electric dryer is rated at 4400 watts when operated from a 220 volt line. What is the resistance of the dryer?

12. Work is equal to the product of power and time. The unit of work sold by electric companies is the kilowatt hour. If one hour = 3600 seconds and 1kilowatt = 1000 watts, how much work does a kilowatt hour represent? (The cost of this amount of work averages about 3 cents.)

13. What current flows in an electric dryer designed to consume 4600 watts from a 220 voltage source?

14. Most houses are wired for a total of 100 amperes to flow. What is the maximum power if all appliances are on 110 volts?

15. Figure 4–26 shows the charge flowing onto a capacitor plate as a function of the time. Estimate the current at the following times:

(a) $t = 0$ seconds,

(b) $t = 1$ second,

(c) $t = 2$ seconds,

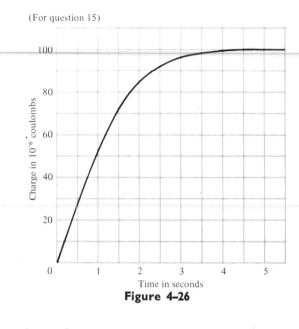

(For question 15)

Figure 4-26

(d) $t = 3$ seconds,

(e) $t = 4$ seconds,

(f) $t = 5$ seconds.

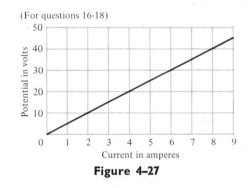

(For questions 16-18)

Figure 4-27

Figure 4–27 shows the variation of current through a resistor with different voltages applied across the resistor.

16. Is the resistance constant?

17. What is the value of the resistance at 3 amperes?

18. How much voltage is required for 5 amperes to flow through the resistance?
Figure 4–28 shows the variation of current through a resistor (a light bulb) with different voltages applied across the resistor.

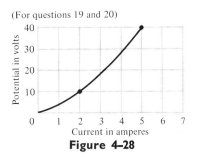

(For questions 19 and 20)

Figure 4–28

19. (a) Is the resistance constant? (b) What is the value of the resistance at 2 amperes?

20. What is the value of the resistance at 5 amperes?

5

MAGNETISM AND ELECTROMAGNETISM

THE NATURE OF MAGNETISM

The ancients knew of rocks (called lodestones) that would attract iron objects. These naturally occurring permanent magnets were used by early mariners as crude compasses. In fact, the words "magnet" and "magnetism" are derived from Magnesia, the region where lodestones were found. Today, permanent magnets are used in many devices, such as refrigerator and cabinet latches and loudspeakers. In addition to permanent magnets, we have electromagnets, which have hundreds of uses. Both permanent and electromagnets can be explained in terms of moving electric charges.

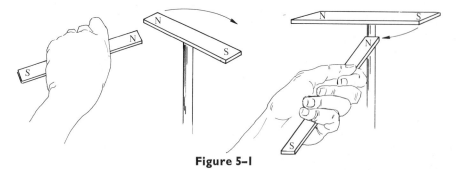

Figure 5–1

A straight bar magnet permitted to turn freely will orient itself in a north-south direction. The magnetic pole that points north is defined as a north-seeking pole and the pole that points south is a south-seeking pole.

It can be found experimentally that *like poles repel each other and unlike poles attract each other*. Two magnetic poles will attract or repel each other in somewhat the same manner as two electric charges or two masses. In fact, of the four forces (gravity, electricity, magnetism, and nuclear) studied by physicists, all but the nuclear (which we do not understand yet) are mathematically similar. That is, the mathematical form is the same. Let's look at a resumé of the three force fields (Table 5–1).

TABLE 5–I

GRAVITY	ELECTRICITY	MAGNETISM
Forces only attractive	Forces attractive for unlike charge but repulsive for like charges	Force attractive for unlike poles but repulsive for like poles
$F = G\dfrac{m_1 m_2}{D^2}$	$F = k\dfrac{Q_1 Q_2}{D^2}$	$F = K\dfrac{p_1 p_2}{D^2}$
F = force in newtons	F = force in newtons	F = force in newtons
m_1 = mass in kilograms	Q_1 = one charge in coulombs	p_1 = one magnetic pole
m_2 = mass in kilograms	Q_2 = other charge in coulombs	p_2 = other magnetic pole
G = constant = 6.67×10^{-11} N-m²/kg²	k = constant = 9.0×10^9 N-m²/C²	K = constant = 10^{-7} W/amp²
D = displacement in meters	D = displacement in meters	D = displacement in meters

Note that the big difference between the gravitational force and the other two forces is that masses produce only attractive forces, whereas electric charges and magnetic poles can produce either attractive or repulsive forces.

To produce the two different masses, the two different electric charges, and the two different magnetic poles, we can do the following:

(1) for masses, we:
    take one mass,

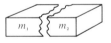

    divide it into two parts,

    and separate m_1 and m_2 by distance D.

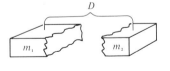

(2) For electric charges, we:
    take a neutrally charged object,

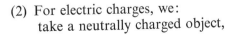

    separate the charge,

Figure 5–2

divide object into two pieces, and

separate the charge by distance D.

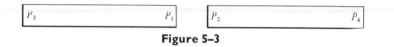

Figure 5–2 *continued*

For magnetic poles, we might try the same scheme, but alas, *it will not work*, because magnetic poles refuse to be separated. We can cut a magnet in half, and we have not two poles, but two magnets (four poles).

P_3 P_1 P_2 P_4

Figure 5–3

We can continue the process of cutting in half, but we cannot succeed in producing a mono-magnetic pole. Therefore, we must put a condition on the magnetic field:

$$F = K\frac{p_1 p_2}{D^2}$$

the condition being that p_1 and p_2 are "isolated" only for mathematical convenience, and the degree of accuracy depends upon how much we can ignore the interaction between p_3 and p_2, between p_1 and p_4, and between p_4 and p_3.

THE MAGNETIC FIELD

We used $\mathbf{g} = $ force/mass to measure the gravitational field and $E = $ force/charge to measure the electrostatic field. We would like an analogous measurement for the magnetic field. We can do this with our isolated magnetic pole. If we take an isolated pole so oriented in a magnetic field that a maximum force is exerted upon it (Figure 5–4), then

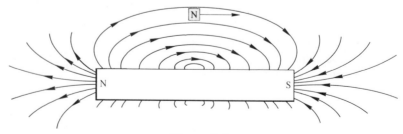

Figure 5–4

we can define the strength of the magnetic induction B^* at any point in space as:

$$B = \frac{F}{p} \quad \text{where } F \text{ is the force in newtons}$$

p is the pole strength in ampere meters

B is the magnetic induction in newtons/ampere meter

As with masses and electric charges, we like to think that a magnet sets up a magnetic field in space. It is also convenient to associate *lines of induction* with the magnetic field. By convention, we say that the direction of an induction line (and therefore, the direction of the magnetic field anywhere in space) is the direction in which the north pole of a compass would point. A compass is nothing more than a magnet that is free to rotate in a plane. If we could devise a freely floating compass, that is, one that could align itself in any direction, the compass would always align itself in the direction of the magnetic field. As in the electric field, if the lines are diverging like those on the right of Figure 5–5, the magnetic field weakens from a to b; if the magnetic lines converge like those on the left of Figure 5–5, the magnetic field strengthens from a to b.

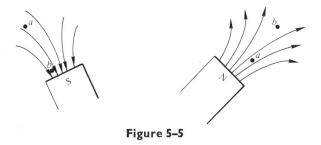

Figure 5–5

Also, the spacing of the lines indicates the magnitude of B: in a weak field, the lines are far apart, and in a strong field the lines are close together. Although the induction lines are continuous, with no beginning or ending, it is convenient to imagine that they emerge from the north pole of a magnet and enter into the south pole (Figure 5–6).

Although the magnetic field cannot be seen, iron filings sprinkled on a piece of cardboard or glass held over a magnet will give a good visual picture of the approximate field of the magnet. The magnetic field is not affected by glass or cardboard, and the iron filings become small magnets that line up in the field. The field can be more accurately described by

---

* Actually, the magnetic induction B is more rigorously defined in units of torque and a magnetic moment vector. We use this restricted definition for simplicity.

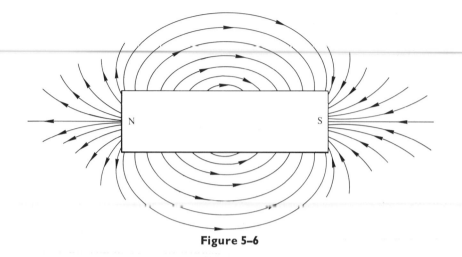

Figure 5–6

placing a small magnet everywhere the field is to be studied and noting the direction of the induction line, as pointed out by the compass. Figure 5–7 shows two different magnetic fields.

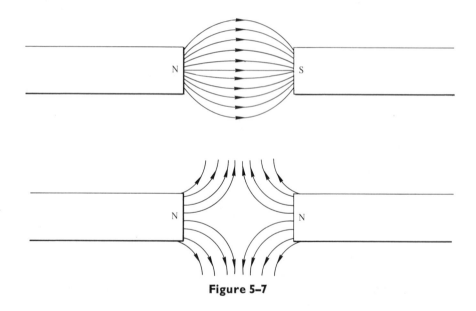

Figure 5–7

Whenever a piece of soft iron is placed in a magnetic field, the magnetic field is disturbed in that the nearby induction lines will turn and appear to be drawn toward and through the iron, and the iron will become a magnet as long as it is in the magnetic field.

We can use the idea that induction lines enter a south pole and emerge from a north pole to establish what polarity any point in the soft iron will have. For example, in Figure 5–8, point *A* on the iron will become a south pole, since the induction line enters the iron at point *A*. Point *B* will

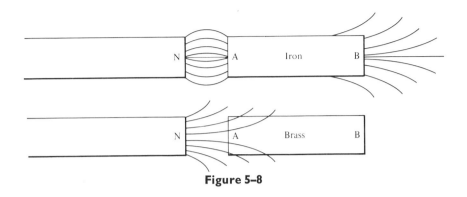

Figure 5–8

become a north pole, since the induction line emerges from point *B*. Notice that the brass does not seem to disturb the field in any way. Only materials that disturb the field (like soft iron) will become magnetized when placed in the field, and such materials are called "ferromagnetic."

Soft iron can also act as a magnetic shield, since lines of force are drawn into the soft iron and iron provides an easier magnetic path than the surrounding material. An anti-magnetic watch uses this principle (Figure 5–9).

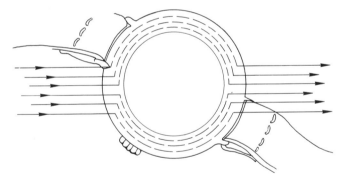

Figure 5–9

PERMANENT MAGNETS

We think of a large magnet as being composed of thousands upon thousands of very small magnets, called magnetic domains, each contributing to the total magnetic field. If the domains are in random directions, the total field will be equal to zero and the material will not show magnetic properties. If a great number of domains are aligned in a given direction, the material will become a strong magnet; in fact, the strength of a magnet depends upon how many of these domains are aligned in the same direction (Figure 5–10).

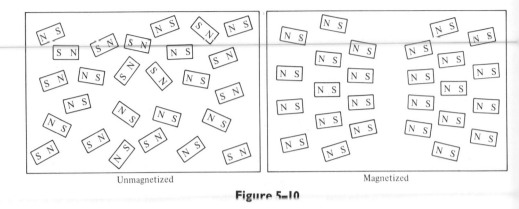

| Unmagnetized | Magnetized |

Figure 5-10

Whenever soft iron is placed in an external field, the domains line up parallel to the field, and when the iron is withdrawn the domains go back to random directions, causing the magnetic properties to cease. If the iron is tempered, however, the domains cannot move freely, and once the iron is magnetized, it will stay magnetized. In order to make a permanent magnet, a bar of tempered steel is heated to give the domains enough thermal energy to rotate freely and is then placed in a strong magnetic field. Once cooled, the bar becomes a strong permanent magnet.

A weak permanent magnet can be made by taking a steel bar (ring stand) aligned with the earth's gravitational field and pounding on the end with a hammer (Figure 5-11). Point *A* will become a south pole (in-

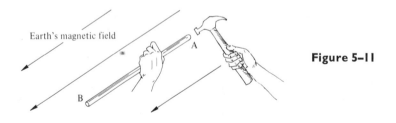

Figure 5-11

duction lines enter) and point *B* will become a north pole (lines emerge). If the bar is placed perpendicular to the field and hammered, it will cease to become a magnet (Figure 5-12).

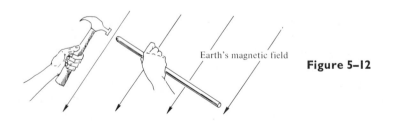

Figure 5-12

An induction line will always give us the direction of the magnetic field anywhere in space. It is convenient to assign a magnitude to the magnetic field induction line to show the strength of the field. To help you visualize this, imagine that a magnet is radiating something out in space that changes space. This "something" we can describe as a magnetic flux, much as a light bulb radiates a light flux. In a space where the flux is dense, the magnetic field is strong, and in a place where the flux is sparse the magnetic field is weak. The question is, just how strong and

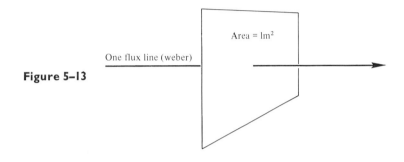

One flux line (weber)

Area = 1m²

Figure 5–13

just how weak? We can give a quantity to this description by defining the magnetic field B in another way: The magnetic field B is equal to the number of induction lines per square meter of surface. B (the magnetic induction) is then equal to the number of flux lines (called webers) emerging perpendicular through this unit area, that is:

$$B = \frac{\phi}{A} \quad \text{where} \quad B \text{ and } A \text{ are perpendicular}$$

ϕ = flux in webers

A = area in meters²

B = webers/meter² and is the magnetic induction and is a measure of the magnitude of the field

We have defined the strength of the magnetic field in two different ways, that is, $B = F/p$ and $B = \phi/A$. By definition, the two must be equal to each other. If force is in newtons, p is in ampere meters, ϕ is in webers, and A is in meters², then by definition:

$$\frac{1 \text{ newton}}{1 \text{ ampere meter}} = \frac{1 \text{ weber}}{\text{meter}^2}.$$

Example 1: At a certain point in a magnetic field, a 2 ampere meter test pole has a maximum force of 10 newtons south exerted upon it. What are the strength and the direction of the magnetic field at this point?

Answer:

$$B = \frac{F}{p} \qquad F = 10 \text{ N south} \qquad p = 2 \text{ ampere meters}$$

$$B = \frac{10 \text{ N}}{2 \text{ amp m}} \text{ south} = \frac{5 \text{ N}}{\text{amp m}} \text{ south, since the direction of}$$

the field is the direction of the maximum force.

Example 2: At a certain point in a magnetic field, there are 100 webers per 5 meters² of area. What is the magnitude of the field?

Answer:

$$B = \frac{\phi}{A} \qquad \phi = 100 \text{ webers} \qquad A = 5 \text{ meter}^2$$

$$B = \frac{100 \text{ Wb}}{5 \text{ m}^2} = \frac{20 \text{ Wb}}{\text{m}^2} \quad \text{or} \quad \frac{20 \text{ N}}{\text{amp} \cdot \text{m}}$$

ELECTROMAGNETISM

If you take a soft iron nail, wrap several turns of insulated wire on it, and connect it to a battery, you will find that the iron nail exhibits all the properties of a permanent magnet as long as charge flows in the wire. What most people do not realize is that if the nail is withdrawn from the

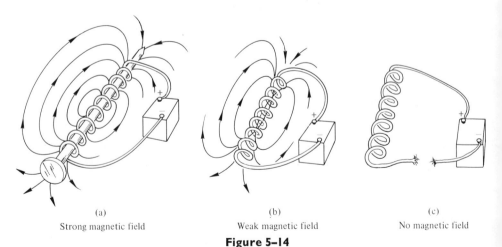

(a)
Strong magnetic field

(b)
Weak magnetic field

(c)
No magnetic field

Figure 5–14

coil, the coil still behaves in every way like a permanent magnet, but with weaker magnetic properties. If the current is cut off, the coil ceases to have magnetic properties. If you take a fully charged capacitor plate, you cannot find any magnetic properties associated with the charge. Therefore, we associate magnetic fields only with *moving* electric charges. Just as a static charge alters space (an electric field), a moving charge produces an additional alteration of space (a magnetic field).

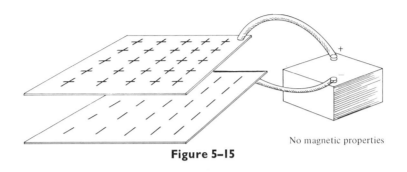

No magnetic properties

Figure 5–15

PARAMAGNETIC, FERROMAGNETIC, AND DIAMAGNETIC MATERIALS

A helically wound coil of wire like that in Figure 5–14a is called a solenoid. If we were to measure the magnetic field B near a current-carrying solenoid and then fill the center of the solenoid with different materials, we would find: (1) an increase in B when the helix is filled with some materials, which are called paramagnetic, ② a great increase in B when the helix is filled with some materials, which are called ferromagnetic, and (3) a decrease in B when the helix is filled with some materials, called diamagnetic materials.

Paramagnetism, ferromagnetism, and diamagnetism can be explained in terms of the electrons of atoms forming current loops, each of which acts as a magnet. In paramagnetic substances, many of the atoms align themselves so as to enhance the external field, whereas in ferromagnetic substances the alignment is much greater and the field is many times stronger than current alone would produce. A *permanent magnet* is produced if the magnetic alignment remains in the material after the current in the helix has stopped. A *temporary magnet* will lose all but a little residual magnetism after the current in the helix drops to zero. In diamagnetic materials, the current loops align themselves in such a way that they oppose the original field and tend to reduce the field set up by the current.

MOVING CHARGES AND MAGNETIC FIELDS

In 1820, Oersted discovered that a moving charge has a magnetic field associated with it. There is a definite relationship between the flow of electrons and the direction of the magnetic field. This relationship is called the "left hand rule." If you pretend that you are holding the electron in your left hand, letting your thumb point in the direction of the velocity of the electron, your fingers will curl in the direction of the induction lines (Figure 5–16).

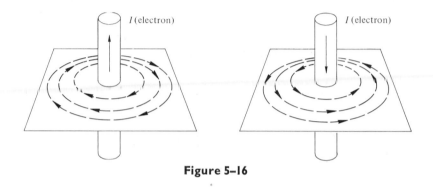

Figure 5–16

It is sometimes convenient to show current or an induction line coming out of the paper. A dot is used to show this because it stands for the head of an arrow. A cross (×) indicates current or an induction line going into the paper. Figure 5–17 illustrates the direction of the induction lines when current is coming out of the paper and going into the paper.

Current out   Current in

Figure 5–17

We can now better understand electromagnetism. If current flows in a circular coil of wire, each charge will have a magnetic field associated with it. By the left hand rule, the field will be as shown in Figure 5–18. The top of the coil will become a north pole and the bottom of the coil will become a south pole if the electron flow is clockwise. The coil behaves in an external magnetic field precisely as a small magnet would (Figure 5–19). Moreover, if we increase the number of turns, increase the current, or increase the area of the coil, the magnetic properties increase. Any permanent magnet that has the same amount of magnetic strength in a given field as a small coil has the same magnetic strength as the coil. The reason for using a coil rather than a permanent magnet is that the

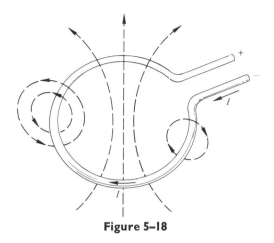

Figure 5–18

magnetic parameters such as current, area, and number of turns of the coil can be calculated much more easily than the parameters associated with a permanent magnet.

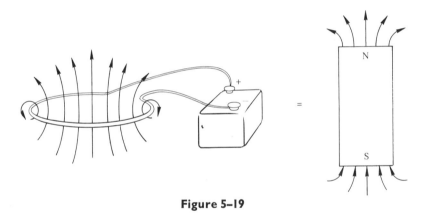

Figure 5–19

Remember that any magnetic field is set up by a moving electric charge. In the case of permanent magnets, the charge is spinning electrons around the nucleus of atoms, causing the "domains"; in electromagnets, the charge is electrons flowing in wires. The direction of the field associated with an electron is given by the left hand rule.

THE MAGNETIC FORCE ON A MOVING CHARGE

Remember the discussion of the oscilloscope in Chapter 4. If both the vertical and horizontal deflecting plates have no voltage, the stream of electrons will make a spot in the middle of the screen like that shown in Figure 5–20.

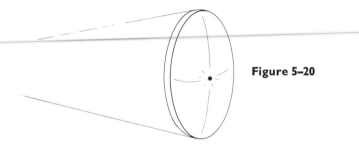

Figure 5–20

If we approach the beam of electrons from above with the north pole of a magnet, the spot will be deflected* as shown in Figure 5–21.

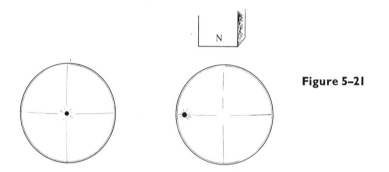

Figure 5–21

Approaching the beam from different directions with a north pole will give the results shown in Figure 5–22.

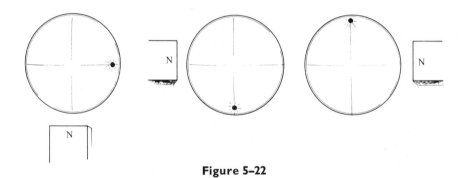

Figure 5–22

Approaching the beam with the south pole of a magnet will also deflect the spot (Figure 5–23).

---

* The electron beam must not, of course, be magnetically shielded.

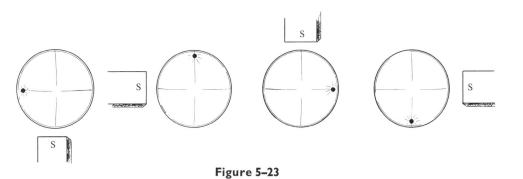

Figure 5-23

From all these observations, we can deduce that the magnetic field of the electron and the external magnetic field due to the magnet interact. Whenever a moving electric charge enters an external magnetic field, the magnetic field associated with the charge and the external field interact. This interaction results in a magnetic force that tends to push the charge sideways. The magnetic induction lines from the two fields will add vectorially and the force on the charge will always be toward the weakest part of the resultant field. For example, imagine an electron coming out of the paper toward you through an external field from a magnet (Figure 5-24).

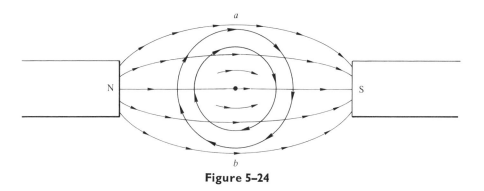

Figure 5-24

It can be seen that the fields set up by the electron (clockwise concentric induction lines) add together to form a stronger field at "*a*" and add to form a weaker field at "*b*." The force on the electron will be downward in this case.*

Newton's law still holds, of course. There is an equal but opposite force on the magnet, but we disregard it.

---

* It is recommended that Experiment 8 in the lab manual be done at this time.

Example: See if you can establish the direction of the force on an electron in the following examples. (Remember to use the left hand rule.)

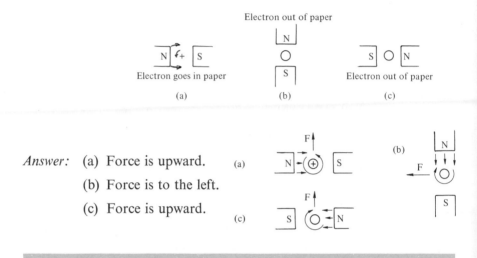

Answer: (a) Force is upward.

(b) Force is to the left.

(c) Force is upward.

It is found experimentally that the force on the charge is proportional to the charge, q, the velocity of the charge, v, and the magnetic induction, B. Only the component of the velocity that is perpendicular to B is used to compute the force, therefore, the relation is:

$$\boxed{F = Bqv}$$ when F is the magnitude of the force in newtons

B is the magnetic induction in $\dfrac{\text{webers}}{\text{meter}^2}$ or $\dfrac{\text{N}}{\text{amp} \cdot \text{m}}$

q is the charge in coulombs

v is the component of the velocity perpendicular to B

Example: What is the force on an electron that moves with a speed of 10^6 m/sec perpendicular to a magnetic field of 0.5 weber/m²?

Answer: $F = Bvq$

$F = ?$

$B = 0.5 \text{ Wb/m}^2 = 0.5 \text{ N/amp} \cdot \text{m}$

$q = 1.6 \times 10^{-19} \text{ coulomb}$

$v = 10^6 \text{ m/sec}$

$$F = \left(\frac{0.5 \text{ N}}{\text{amp m}}\right)(10^6 \text{ m/sec})(1.6 \times 10^{-19} \text{ C})$$

$$F = 8 \times 10^{-14} \text{ newton}$$

The direction of the force will be perpendicular to the velocity (toward the weaker field) and will tend to cause the electron to travel in a circle.

THE CYCLOTRON

You will learn in Chapter 10 that man has probed deeply into the secrets of the atom. We have sent very energetic charged particles into the atom to see what comes out. One of the several machines that will give large energies to the charged particle is the cyclotron. A cyclotron does not give all the energy to a charged particle at one time, but it gives the particle the energy in steps. We use an electrostatic field to give the energy to the particle and use a constant magnetic field to turn the particle between each energy "step."

A cyclotron consists of two hollow metal plates (called dees) that will *not* shield a magnetic field and a large magnet to supply a constant magnetic field. Figure 5–25 is a schematic diagram of a cyclotron.

A charged particle such as a proton enters the cyclotron in the middle and between the dees. The voltage between the dees accelerates the charge while the charge travels the space between the dees. When the charge is inside the hollow metal dee, there is no electrostatic force on the charge

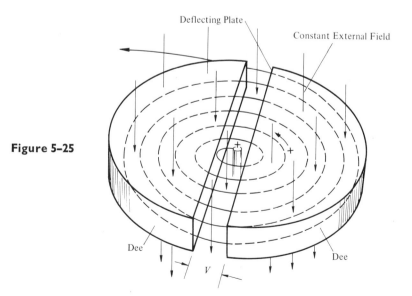

Figure 5–25

Deflecting Plate

Constant External Field

Dee

Dee

V

because the metal effectively shields the electrostatic force. However, the magnetic field goes through copper as well as through air, so the charge will always have a magnetic centripetal force on it.

During the time that the charge travels the circular path in one of the dees, the voltage across the dees is reversed, and the charge is accelerated each time it travels between the dees. Since the charge is accelerated each time it enters the space between the dees, the charge will travel in ever widening circles, making a spiral path. The energy given to the particle is limited only by the radius of the circle until the speed of the particle approaches the speed of light. When the particle approaches the speed of light, the mass of the particle begins to increase (Chapter 8) and the time the charge spends turning around inside the dees gets out of step with the voltage across the dees. When this happens, the voltage between the dees is just as likely to subtract energy from the particle as to add energy to it.

Physicists have devised many atomic accelerators, such as the synchrocyclotrons and synchrotrons, to alleviate the economic problem of building large magnets and to alleviate the relativistic limitations. Accelerators of this type can accelerate protons to speeds of 0.99998 the speed of light. At this speed, the proton has been accelerated through approximately 30 billion volts!

THE ELECTRIC GENERATOR

Electric power is one of the major energy sources in the world today. The energy to run electric generators is derived from the energy of falling water (hydroelectric power), from combustible fuel (coal and oil), and from nuclear fission.

Although a commercial generator is somewhat complex, we can understand the principles involved by using a very simple model generator. The model generator consists of a U-shaped conductor. A conducting bar slides upon it, keeping electrical contact with the U-shaped bar. As such, our generator is nothing more than an expandable conducting loop. An essential part of a generator is an "external" magnetic field.

Imagine that the ceiling is the north pole of a magnet and that the floor is the south pole of a magnet. We would then have an external field going into this paper (Figure 5–26). Next, take the straight metal con-

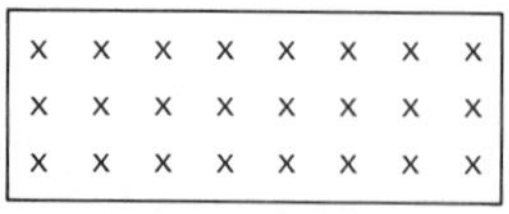

Figure 5–26

ductor bar and move it to the right with a velocity v across a U-shaped conductor. Each electron in the straight metal rod would then be moving with a velocity v across the paper. A moving charge sets up a magnetic

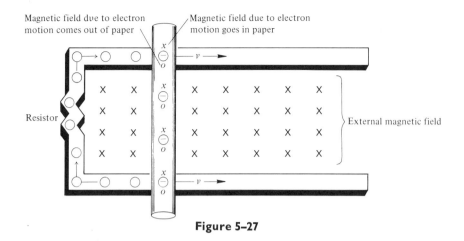

Figure 5–27

field. The field associated with each electron would be given by the left hand rule (Figure 5–27).

The external field and the field due to the electron would add to give a stronger field on the side of the electron next to the top of the page. The fields would add to give a weaker field on the side of each electron next to the bottom of the page. Therefore, each electron would experience a force pushing it toward the bottom of the straight conductor. Since charges will not pile up, the magnetic force on the electrons in the straight conductor affects every electron in the circuit, causing a current to move clockwise, as indicated. We use a magnetic force to work on electrons and push them through a conducting circuit like the U-shaped bar. Work is done *on the electrons* (work/charge) in the straight conductor, which causes a *potential or voltage rise* (called electromotive force, or emf) across the straight conductor. The work is dissipated by electrons working on the resistance of the circuit. Resistances are called potential sinks or voltage drops. Experimentally, it is found that:

1. Electrons will flow clockwise as long as the straight conductor is pushed to the right.
2. Electrons will flow counterclockwise if the straight conductor is pulled back to the left.
3. Electrons will stop flowing when the straight conductor stops.
4. Electrons will stop flowing if the external field is removed.
5. The larger the electron flow, the larger the force required to push the bar.
6. The magnetic field set up by the induced current in the loop will always be such that it tends to prevent movement of the bar.

In addition, the flow of electrons will be increased if the velocity of the straight conductor is increased or the amount of magnetic flux is increased. In the absence of any frictional forces, the work the magnetic field does on the electrons is precisely equal to the work we do against the magnetic field. In order to understand this more completely, let's go through what happens in the U-bar generator, step by step.

STEP 1:   We push the bar to the right with a mechanical force F_m (Figure 5–28). The work we do will be the force times displacement.

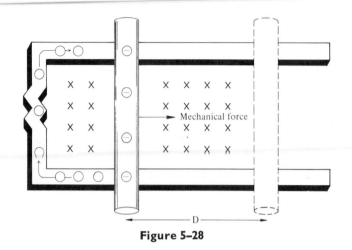

Figure 5–28

STEP 2:   As the bar moves perpendicular to the external magnetic field with a velocity v, a magnetic force (parallel to the bar) is on the electrons in the bar. This force is equal to Bvq, where B is the external field, v is the velocity of the bar to the right, and q is the charge flowing

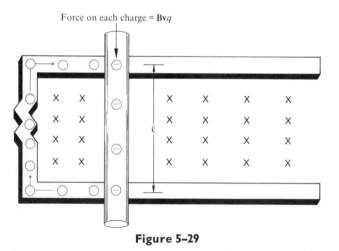

Figure 5–29

through the bar (Figure 5–29). The magnetic force is parallel to the length ℓ of the bar, so the work the magnetic force does on the charge as it travels the length of the bar is:

Work = (magnetic force)(length of bar)

work $= (Bvq)(\ell)$; the voltage across the bar is $V = \dfrac{\text{work}}{\text{charge}}$

$$or \quad \frac{W}{q} = \frac{(Bvq)}{q}(\ell)$$

$$V = Bv\ell$$

STEP 3: The magnetic force on the charges in the bar causes a current to flow clockwise. Since current will be flowing in the bar, a magnetic field is set up (Figure 5–30). The field set up by the electron flow comes out of the paper on the left side of the bar and goes into the paper on the right side of the bar.

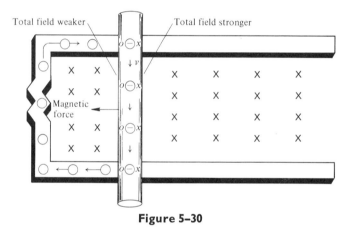

Figure 5–30

However, the external field goes into the paper; therefore, the total field will be weakened on the left side of the bar, and each electron will also have a magnetic force exerted on it to the left. This magnetic force is equal to but is in the opposite direction from the mechanical force that keeps the bar traveling with a velocity v.

The preceding statement is simply a conservation of energy principle known as *Lenz's Law*. *The direction of an induced current will always be such that it will oppose the cause that set it up.*

Neglecting frictional forces, the mechanical work that we do on the bar will be equal to the work done on the electrons by the magnetic forces, which is equal to the work the electrons do on the circuit elements. That is:

$$FD \quad = \quad (Bvq)(\ell) \quad = \quad VIt$$

mechanical work	= work done on	= work extracted
done against	electrons by	from electron by
magnetic field	magnetic field	resistances.

As a numerical example, suppose that the external B field is 5 N/amp · m, the velocity of the bar is 20 m/sec, the length of the bar is 1 meter, and the circuit element has a resistance of 5 ohms. We would like to know the emf of the generator, the current through the resistance, the power consumed by the resistance, and the mechanical force we must exert on the bar to maintain the velocity of 20 meters/sec.

First, the emf of the generator can be obtained from:

$$V = Bv\ell$$

$$V = Bv\ell = (5 \text{ N/amp} \cdot \text{m})(20 \text{ m/sec})(1 \text{ m}) = 100 \text{ volts}.$$

The current through the resistor can be obtained using Ohm's Law:

$$I = \frac{V}{R} = \frac{100 \text{ volts}}{5 \text{ }\Omega} = 20 \text{ amperes}.$$

Since we know the voltage across and the current through the resistor, the power is:

$$P = (100 \text{ volts})(20 \text{ amp}) = 2000 \text{ watts}.$$

The work done each second is 2000 joules, therefore:

$$F \cdot D = 2000 \text{ joules},$$

and since the bar travels with a velocity of 20 meters/sec, the displacement is 20 meters each second.

$$F \cdot 20 \text{ m} = 2000 \text{ joules}$$

$$F = 100 \text{ newtons}.$$

A force of 100 newtons must be applied to overcome the force of the magnetic field against the bar.

Notice that if the circuit is broken, no power is consumed, and the force necessary to maintain the bar at a velocity of 20 m/sec would be equal to zero. The greater the power consumption, the greater the mechanical force to keep the bar traveling at 20 m/sec. In electrical generating plants, the rotational speed of the generator is constant, and every time an appliance is turned on, the force necessary to maintain the rotational speed of the generator is increased. The increased force is obtained from a greater amount of steam or a greater amount of water flowing through the turbine blades that drive the generator.

Whenever the bar travels through the magnetic induction lines, the bar is said to "cut" magnetic lines. Now, since all velocities are relative, we wonder if, rather than pushing the bar to "cut" magnetic induction lines, we could keep the bar stationary and move or change the induction lines. We can derive one of the most fundamental laws of electromagnetic

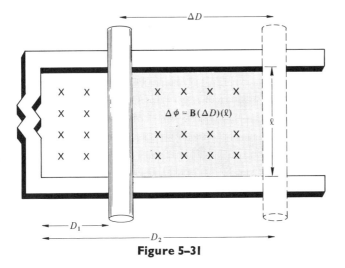

$$\Delta \phi = B(\Delta D)(\ell)$$

Figure 5–31

induction using our model generator and the formula for the voltage across the bar, $V = Bv\ell$.

When the bar is at D_1, the bar has cut the amount of flux lines ϕ_1, and when the bar is at position D_2, the bar has cut flux lines equal to ϕ_2. The change in the amount of flux through the loop, $\Delta\phi$, is the flux between the two positions. This change in the flux is equal to the magnetic field B times the change in the area, since by definition the magnetic field $B = \phi/A$. The change in the area is the shaded portion between the two bar positions (a rectangle) and is equal to the change in the displacement of the bar multiplied by the length of the bar, that is:

$$\Delta A = \Delta D\ell$$

Now $V = Bv\ell$ or, since the velocity v by definition is $\Delta D/\Delta t$,

$$V = \frac{B(\Delta D\ell)}{\Delta t} \quad \text{or} \quad V = \frac{B\,\Delta A}{\Delta t}\ .$$

But since $B\,\Delta A$ is the change in flux, $\Delta\phi$:

$$V = \frac{\Delta\phi}{\Delta t}\ .$$

From Lenz's Law, we know that the induced voltage will oppose the agent that set it up, so we add a minus sign to remind us of this fact:

$$V = -\frac{\Delta\phi}{\Delta t}\ .$$

The foregoing is known as Faraday's Law of Electromagnetic Induction. We have shown that it does not matter whether a conductor cuts a magnetic field or a magnetic field cuts a conductor; an emf will be induced in a conducting loop in either case. Whenever the flux encircled by a conducting loop changes, a potential rise will be induced in the loop.

We can do an experiment to test all of these ideas. Take a coil of wire, a sensitive ammeter, and a magnet, and do a series of experiments (Figure 5–32). As the magnet is moved toward the coil, a current is set up in the coil. The direction of the current will be to oppose the motion. This means that the end of the coil facing the magnet must become a north pole, since like poles repel. By the left hand rule, this can only happen if the charges are flowing clockwise in the coil. If the coil is brought to the magnet, the results will be the same. If the relative motion stops, the current in the coil will drop to zero.

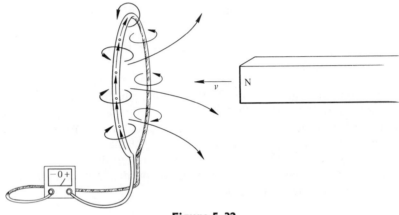

Figure 5–32

If the direction of the magnet is reversed (Figure 5–33), the current in the coil will reverse direction to oppose the motion of the magnet, since unlike poles attract each other. If we now use a south pole and approach

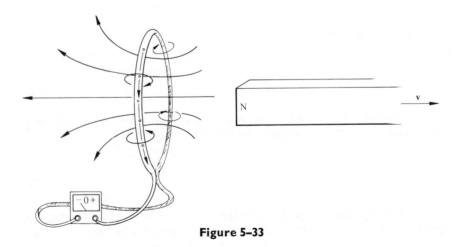

Figure 5–33

the coil (Figure 5–34), the side of the coil facing the magnet will become a south pole to oppose the motion. By the left hand rule, this can only happen if the electrons are flowing counterclockwise. Again, if the motion is stopped, the current in the coil drops to zero, and if the motion of the magnet is reversed, the polarity and direction of the current in the coil reverse. If this were not true, we could have a perpetual motion machine. We could start the magnet toward the coil, and if the coil would attract the magnet, we could have the current doing work and at the same time accelerating the magnet. This "something for nothing" deal never happens in nature.

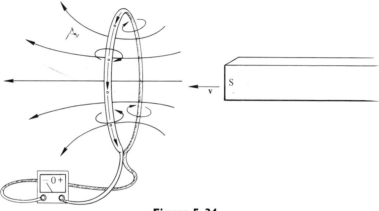

Figure 5–34

We will now replace the magnet with another coil of wire that is connected to a variable current supply (Figure 5–35). Since a coil of wire with a current through it is an electromagnet, we can change the magnetic flux by changing the current in the electromagnet. We will keep the coils

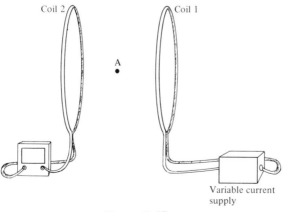

Figure 5–35

a given distance from each other. If the electron flow in coil 1 is clockwise, the side facing coil 2—that is, side *A*—will become a south pole. If we now increase the current in coil 1, side *A* will become a stronger south pole. The procedure would be analogous to approaching coil 2 with the south pole of a permanent magnet. We already know that in this case the current in coil 2 will be counterclockwise. If we decrease the clockwise flow of current in coil 1, side *A* becomes a weaker south pole. It would be analogous to moving the south pole of a magnet away from coil 2. If we let a steady current flow in coil 1, the current in coil 2 drops to zero, analogous to having no relative motion between coil 2 and a permanent magnet. If we oscillate the current in coil 1, the current in coil 2 will also oscillate, and will always be in such a direction as to oppose any change in the current of coil 1. The amount of emf in coil 2 depends only upon the rate of change of the flux that is enclosed by the coil.

Therefore, in the final analysis, we can state: Whenever a magnetic flux is changing in a conducting loop, a voltage rise takes place in the conductor. That is:

$$\text{Potential rise (in volts)} = -\frac{\Delta\phi \text{ (in webers)}}{\Delta t \text{ (in seconds)}} \cdot *$$

* It is recommended that Experiment 9 in the lab manual be done at this time.

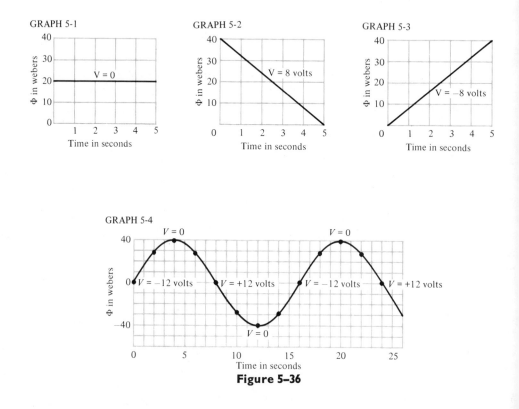

Figure 5–36

If the flux is rising, there is a voltage in one direction; if the flux is falling, there is a voltage in the opposite direction. The voltage at any particular time would be the negative slope of a ϕ versus time curve. The graphs in Figure 5–36 illustrate.

The voltage induced will always be in such a direction that the current through the circuit will set up a flux that opposes any change in the original flux that is causing the voltage.

ALTERNATING VOLTAGE

The U-shaped conductor with the straight conductor sliding across it is a very simple electric generator. As long as the magnetic field, the velocity of the bar, and the length of the bar are constant, the voltage output will be constant, and we say we have a direct current generator. However, in order to keep the voltage constant, we must have a constant external field, and this is impossible to maintain over long distances. Also, we would have to keep pushing the straight bar in the same direction forever, which would prove to be quite a chore. If we stop the straight conductor and come back to the right, the potential difference or emf will change directions. If we push the straight rod back and forth in a regular manner, the voltage will oscillate in the same manner, and we have what is called an *alternating voltage*. The most convenient method to generate a voltage is to have the conductor rotate in a circle, which causes the voltage to oscillate from positive to negative (Figure 5–37).

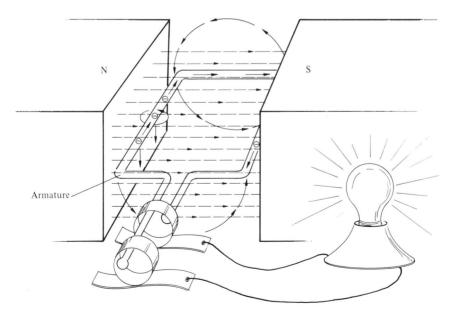

Armature

Figure 5–37

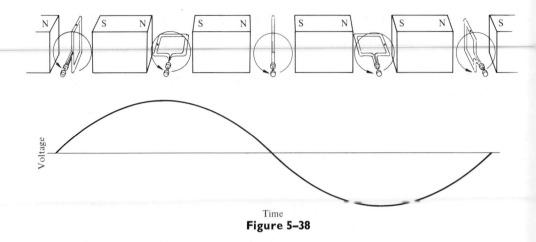

Time

Figure 5–38

At some positions, the armature is cutting a maximum amount of flux per unit of time ($\Delta\phi/\Delta t$ = maximum), and at other positions it will be cutting no flux per unit of time ($\Delta\phi/\Delta t$ = 0). The different positions of the armature are shown in Figure 5–38, along with the voltage output at each particular position. The curve is called a sine curve. Across a conductor, the current equals voltage divided by the resistance, which is constant; therefore, the current alternates and is the same type of curve as the voltage.

POWER IN AN ALTERNATING CIRCUIT

The power dissipated at any instant by a resistor is the product of the potential difference across the resistor and the current through the resistor. Both the voltage drop across the resistor and the current through the resistor are plotted on Graph 5–5. If the voltage alternates, so will the

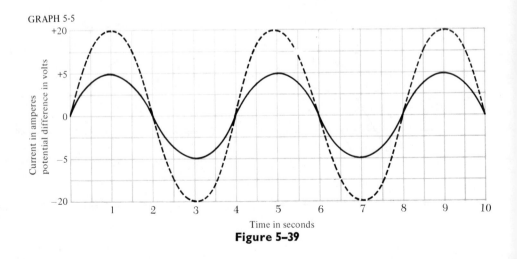

GRAPH 5-5

Time in seconds

Figure 5–39

GRAPH 5-6

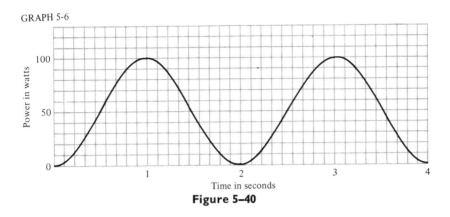

Figure 5–40

current (the dotted line), and since the value of the voltage and the current will be both positive and both negative at any given instant, the power will always be positive. If we were to plot the product of the curves in Graph 5–5 over one complete revolution of the generator, we would get a curve like that in Graph 5–6. Note that the maximum power is (5 volts) × (20 amperes) = 100 watts and that the minimum power is zero. The power produced by the electric power companies alternates 60 times each second. It is necessary to know the amount of work a circuit will do if the power is oscillating as in Graph 5–6. Since the power varies with time, we must compute the work by finding the area under a power versus time curve (Graph 5–7).

GRAPH 5-7

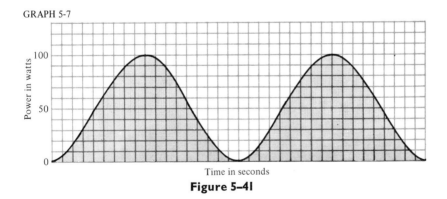

Figure 5–41

Finding the area under the curve is a lot easier than it looks. Let's draw a dotted line half-way up the "peak" value of the power (Graph 5–8). The peak power is 100 watts "high," so the dotted line is drawn 50 watts high. We can easily see that the shaded area of the peaks that are above the dotted line will just fill in the area of the "troughs" that are below the dotted line. It is easy to see that the "average" or "effective" power is just one half of the peak power, that is,

average power = ½ peak power.

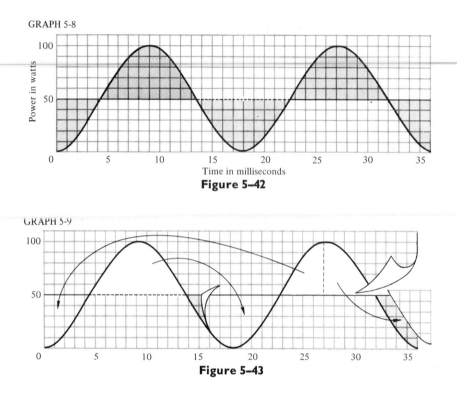

GRAPH 5-8

Power in watts

Time in milliseconds

Figure 5-42

GRAPH 5-9

Figure 5-43

Now that we have found that the average power is half the peak power, we would like a method for finding it directly from the product of the potential difference and the current rather than having to multiply the voltage and the current point by point and plotting to find the power. We can do this by reducing both the peak voltage and the peak current by a given percentage. In order to find how much we must reduce each, we start with the relation:

$$\tfrac{1}{2}\text{ peak power} = \text{average power.}$$

The peak power is the product of the peak potential difference and the peak current, so we can write:

$$\tfrac{1}{2}(V_pI_p) = \text{average power.}$$

We must reduce both the peak potential difference and the peak current by a factor so that the product of the two will be reduced by one half. Also, the two must be reduced by the same amount, so what two equal factors when multiplied together will give us one half? The answer is easy—it must be the square root of one half. Therefore:

$$\frac{V_p}{\sqrt{2}}\frac{I_p}{\sqrt{2}} = \text{average power}$$

$\dfrac{V_p}{\sqrt{2}}$ is defined as the *effective* potential difference and

$\dfrac{I_p}{\sqrt{2}}$ is defined as the *effective* current

$$\dfrac{V_p}{\sqrt{2}} = V_{(\text{effective})}$$

$$\dfrac{I_p}{\sqrt{2}} = I_{(\text{effective})}.$$

In alternating circuits, *all potential differences and all currents are understood to be effective unless we state otherwise.* For example, the voltage of 110 volts AC is effective voltage and the peak voltage is $(\sqrt{2})(110)$, or approximately 155 volts. Most voltmeters and alternating current ammeters are manufactured to read effective voltage and effective current, and if peak values are needed they can be obtained from the relationships just given.

Example 1:   An insulator can withstand a voltage of 1410 peak volts before breakdown. What is the effective voltage?

Answer: $V_{(\text{eff})} = 0.707\, V_{\text{peak}} = 1000$ volts.

GRAPH 5-10

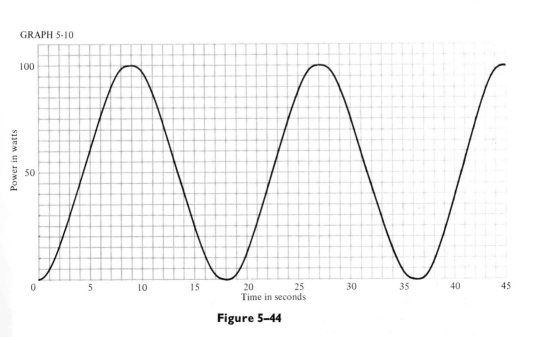

Figure 5–44

The following problems should give you some insight into the concepts of power just discussed. Figure 5–44 shows the power across a 25 ohm resistor as a function of time.

Example 2:   What was the peak power?

   Answer: From the graph, power = 100 watts.

Example 3:   At what time(s) was the power 59 watts?

   Answer: From the graph, $t = 5$, 13, 23, 31, and 41 seconds.

Example 4:   What was the average power?

   Answer: Average power = ½ peak power = ½(100 watts) = 50 watts.

Example 5:   Using the relation $P_{peak} = I_{peak}^2 R$, find the peak current.

   Answer: $100 = I^2(25)$

$$I^2 = 4 \quad or \quad I = 2 \text{ amperes.}$$

Example 6:   Using Ohm's Law, $V = IR$, find the peak voltage.

   Answer: Peak $V = I_{peak}R$; $V = 2 \cdot 25 = 50$ volts.

Example 7:   What is the effective current?

   Answer:

$$I_{(eff)} = \frac{I_p}{\sqrt{2}} = \frac{2 \text{ ampere}}{\sqrt{2}} \approx 1.4 \text{ amperes.}$$

Example 8:   What is the effective potential difference?

   Answer:

$$V_{eff} = \frac{V_p}{\sqrt{2}} = \frac{50 \text{ volts}}{\sqrt{2}} \approx 38 \text{ volts.}$$

THE TRANSFORMER

Now that you have studied the generator and Faraday's Law, $V = -(\Delta\phi/\Delta t)$, you are in a position to understand how a transformer works. In the simplest sense, a transformer can be thought of as a generator with no moving parts. The only movement in a transformer is the changing magnetic field and the movement of electrons.

A transformer consists essentially of three parts, the primary winding,

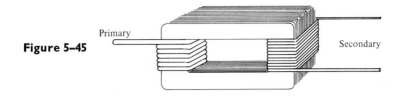

Figure 5–45

a ferromagnetic core (such as soft iron or silicon steel), and a secondary winding (Figure 5–45). The function of the primary winding is to set up a changing magnetic flux when it is connected to an alternating current source. The ferromagnetic core guides the flux through the secondary coil and the changing flux sets up an alternating voltage across the secondary coil. Each winding on the primary will create a given amount of magnetic flux for a given voltage, and each winding on the secondary will encircle this magnetic flux. Therefore, if there are more windings on the primary than on the secondary, there will be less voltage on the secondary, and this is called a step-down transformer. On the other hand, if there are more windings on the secondary, the voltage will be greater on the secondary, and it is called a step-up transformer. The core of a transformer is usually made from strips of iron held together (laminated) in order to stop "eddy" currents in the core. Eddy currents only serve to heat the core and lower efficiency. Most transformers are very efficient, so the following relations apply:

$$\frac{\text{Primary turns}}{\text{Secondary turns}} = \frac{\text{voltage across primary}}{\text{voltage across secondary}}$$

and

$$\frac{\overbrace{\text{Power in}}}{V_{(\text{primary})}\, I_{(\text{primary})}} = \frac{\overbrace{\text{power out}}}{V_{(\text{secondary})}\, I_{(\text{secondary})}}.$$

Without transformers, the efficient transmission of electrical power would be impossible.

In order to minimize power line losses, electric companies step up the voltage to very high values (220,000 volts) at the power station and step-down the voltage (to around 12,200) at power substations located in the vicinity where the electricity will be used. A further reduction of voltage occurs (±110) just before the power is used. (Note the transformers located on poles throughout your neighborhood.)

Example: It is desired to step up the voltage from 110 to 11,000 volts. (a) If the primary has 1000 turns, how many turns must there be on the secondary? (b) What is the current in the secondary when the current in the primary is 50 amperes?

Answers: (a) $\dfrac{\text{Turn}_p}{\text{Turn}_s} = \dfrac{\text{Voltage}_p}{\text{Voltage}_s}$

$$\frac{1000 \text{ turns}}{x} = \frac{110 \text{ volts}}{11,000 \text{ volts}}$$

$$110x = (1000)(11,000) \text{ turns} = 100,000$$
$$\text{turns.}$$

(b) $V_p I_p = V_s I_s$

$$(110 \text{ volts})(50 \text{ amperes}) = (11,000 \text{ volts})I_s$$

$$I_s = \frac{(110)(50) \text{ amp}}{11,000} = \tfrac{1}{2} \text{ ampere.}$$

EXERCISES

1. A small magnet of pole strength 2 ampere · meter experiences a force of 4 newtons north at a point in space. What is the magnetic induction at that point?

2. The magnetic induction at a certain spot is 10 N/amp · m. If a magnetic pole of 2 ampere · meter is placed at this spot, what is the force on the magnet?

3. What similarities are there in the gravitational, electrostatic, and magnetic forces? How do the forces differ from each other?

4. Suppose you are at a distant point in space. Tell how you could devise an experiment to find out whether or not there were: (a) a gravitational field, (b) an electrostatic field, and (c) a magnetic field.

5.
$$\boxed{N \quad S} \leftarrow d \rightarrow \boxed{A}$$
iron

(a) What type of magnetic pole is point A? (b) If the distance d is increased, what happens to the strength of A?

6. How do paramagnetic, ferromagnetic, and diamagnetic materials differ?

7. If 10 flux lines (webers) emerge from a solenoid of 0.2 meter^2 in area, what is the magnetic induction in the solenoid?

8. The magnetic induction is 5 webers/m^2 at a certain point in space. What is the minimum area that would contain 10 webers?

9. If an electron is coming out of the page toward you, what is the direction of the induction lines?

10. A conductor encircles a magnetic field of 50 Wb/m^2. What is the emf in the conductor if the field drops to 0 in 2 seconds?

11. Assume that the magnetic field of the Earth is vertical $\downarrow\downarrow$. Describe what would happen if you played "skip the rope" with a conductor hooked to a voltmeter and you turn the rope clockwise (Figure 5–46).

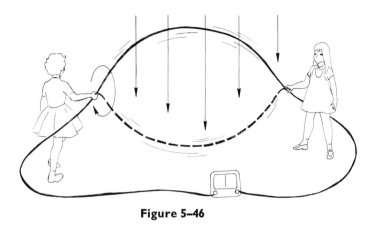

Figure 5–46

Figure 5–47 shows the flux through a coil as a function of time. Answer the following questions.

12. What is the flux at $t = 5$ seconds?

13. Is the emf constant?

14. What is the magnitude of the voltage?

(For questions 12-14)

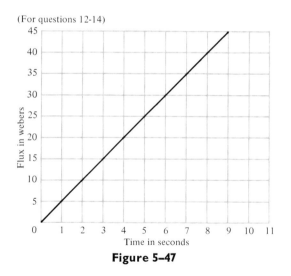

Figure 5–47

Figure 5–48 shows the flux through a coil as a function of time. Answer the following questions.

15. What is the flux at $t = 5$ seconds?

16. What is the induced emf at $t = 3$ seconds?

17. What is the induced emf at $t = 7$ seconds?

(For questions 15-17)

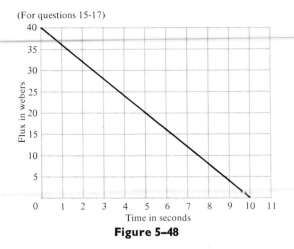

Figure 5-48

18. A charge of 10^3 microcoulombs enters a uniform magnetic field of 5 Wb/m² traveling at a speed of 10^3 meters/sec. What is the force on the charge?

19. What is the force on an electron traveling 10^5 m/sec that enters a perpendicular magnetic field of 2 Wb/m²?

20. In Problem 19, what is the radius of the circle that the electron would travel around if the mass of the electron were 9×10^{-31} kg?

(For questions 21-25)

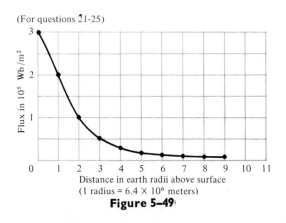

Figure 5-49

An electron from outer space enters the Earth's magnetic field. The velocity of the electron remains constant, but the magnitude of the magnetic field B increases according to the graph. Answer the following questions.

21. Is the magnetic flux increasing or decreasing as the electrons near the surface of the Earth?

22. Is the force on the electron increasing or decreasing?

23. Plot a graph of the force on the electron, assuming that the charge on the electron is 1.6×10^{-19} coulombs and that the velocity is 10^5 meters/sec.

24. What would be the centripetal acceleration of the electron ($\mathbf{F} = m\mathbf{a}$) at $d = 3$ radii?

25. What would be the radius of the circle that the electron would tend to go around at $d = 3$ radii
($\mathbf{F} = m\mathbf{a} = mv^2/R$)?

Figure 5–50 shows the flux through a loop as a function of time. Answer the following questions.

26. At what time is the voltage a positive minimum?

27. At what time is the voltage a positive maximum?

28. What is the voltage at $t = 5$ seconds?

29. At what time is the voltage zero?

30. What is the maximum negative voltage?

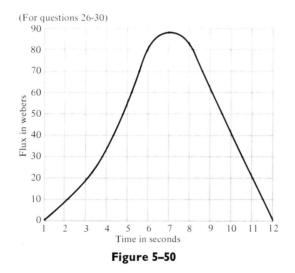

(For questions 26-30)

Flux in webers — Time in seconds

Figure 5–50

Figure 5–51 shows the flux cut by a conducting loop as a function of time. Answer the following questions.

31. At what time(s) was the flux a maximum?

32. At what time(s) was the induced voltage a maximum?

33. At what time(s) was the induced voltage zero?

34. What type of voltage is being generated?

(For questions 31-34)

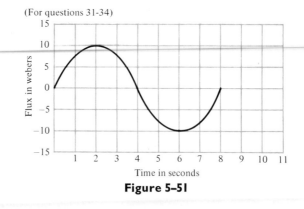

Figure 5–51

35. If an electron is coming out of the paper toward you and the external magnetic field is as illustrated, find the direction of the force on the electron.

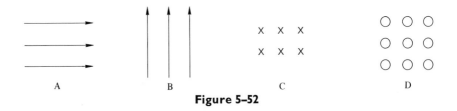

Figure 5–52

36. It is desired to force an electron that is traveling in the direction shown (Figure 5–53) to be deflected downward. Where would you place the north pole of a magnet to do this?

Figure 5–53

37. How could the voltage of a generator capable of inducing 100 volts in a current loop be doubled?

38. It takes a force of 5 newtons to turn a hand generator. If the generator is connected to a light bulb, will the force increase, decrease, or remain the same?
    Figure 5–54 shows the electron flow through a U-shaped generator as a function of the velocity. Answer the following questions.

39. If we wish to generate a current of 20 amperes, what should be the velocity of the straight conductor?

40. If the total resistance of the electron path is 10 ohms, what is the emf between the ends of the straight conductor when it is traveling (a) 10 meters/sec? (b) 30 m/sec?

(For questions 39 and 40)

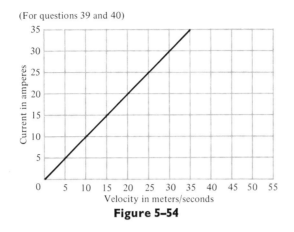

Figure 5–54

Figure 5–55 shows the power across a 5 ohm resistor as a function of time. Answer the following questions.

41. What was the peak power?

42. At what time(s) was the power 10 watts?

43. What was the average or effective power?

44. How much actual work was done in 45 seconds?

45. Using the relation $P_{peak} = I_{peak}^2 R$, find the peak current.

46. Using the relation $P_{(effective)} = I_{eff}^2 R$, find the effective current.

47. Using Ohm's Law, $V = IR$, find (a) the peak voltage and (b) the effective voltage.

(For questions 41-47)

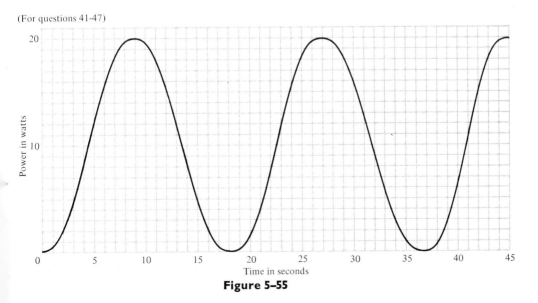

Figure 5–55

48. An insulator can withstand a potential of 2820 volts before breakdown. What is the effective voltage?

49. A transformer has 100 turns on the primary and 10,000 turns on the secondary. What is the emf across the secondary when 200 volts are across the primary?

50. A transformer is needed to supply 11,000 volts from a 110 volt supply. What must be the turns on the primary if there are 5000 turns on the secondary?

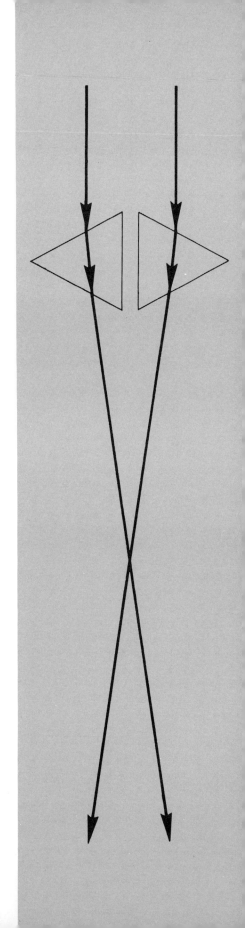

WAVES

6

THE WAY OF WAVES

Thus far in this course, we have considered the description of the physical world in terms of such concepts as motion, force, and energy. Our understanding of each of these concepts has involved relating the concept to matter—like rockets, springs, and particles. We now ask, "Is there anything except particles that can move from one point to another? Is there any way to transmit energy that does not involve the transportation of matter?"

If you stand at the end of a bowling alley and you wish to cause the pins at the other end to fall, the most enjoyable and one of the most practical ways of accomplishing this is by rolling a collection of particles down the alley. But could the pins be knocked down without any particles moving from one end of the alley to the other? Yes, a very loud sound can level the pins, without the transportation of matter. Obviously, work has been done on the pins; therefore, energy has been transmitted from one end of the alley to the other. If no particles have been transported, what is it that has moved the length of the alley? The answer is simply a *disturbance*. A disturbance that carries energy from one place to another is called a *wave*. Since there are many types of disturbances (water waves, sound, light, and so on), a description of the physical universe would not be complete without an understanding of the properties of waves.

THE WAVE CONCEPT

One of the simplest ways to visualize waves and thus study their properties is through the observation of waves on a coiled spring. Observe carefully the disturbance produced and transmitted on a Slinky (Figure 6–1). Notice that the source of the disturbance is the to-and-fro motion of the hand. This means that the wave is being produced by a vibration or

167

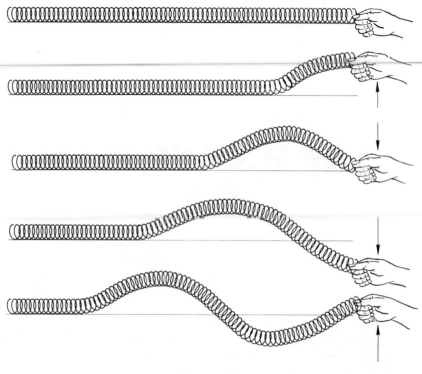

Figure 6–1

oscillation and that there is one wave per vibration. Now observe carefully the manner in which this disturbance is propagated by the spring. The disturbance moves from one end of the Slinky to the other, but after the wave has moved past a given coil, that coil is exactly where it was before the disturbance passed. Therefore, energy has been transmitted from one end of the spring to the other with no net transfer of matter in the process. What has moved from one end of the spring to the other is energy in the form of nothing more than a pattern on the spring.

Observe that the vibrations can occur in two basically different ways. When the hand is moved to and fro perpendicular to the spring, the coils of the spring also move at right angles to the direction of energy propagation which is along the spring (Figure 6–2). Such a wave is called a *transverse wave*. However, if the driving mechanism moves parallel to the direction of the spring, so do the coils, and so does the energy. This type

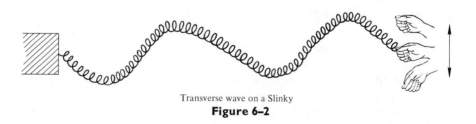

Transverse wave on a Slinky
Figure 6–2

Longitudinal wave on a Slinky
Figure 6–3

of energy propagation in which the vibrations are in the direction of energy transfer is called a *longitudinal wave* (Figure 6–3).

Both types of waves have certain properties in common. Let's attempt to discover some of these properties by making some simple measurements. Position yourself at some point along the spring and observe the waves passing. One characteristic of the passing waves is the time needed for one complete disturbance to pass your position. This time is defined as the *period* of the wave. Suppose that instead you count the number of waves that pass in a given time. This number of complete waves passing a point per unit time is the *frequency* of the wave. Note that period and frequency are not independent. Since period is the time for one wave to pass and frequency is the number of waves that pass in a given time, frequency and period are reciprocals of each other. Another quantity of interest is the length of the disturbance, which is the *wavelength*. Wavelength is measured between two successive points that are executing the same motion in a wave, such as between one crest and the next crest. *Amplitude* is a term used for the maximum displacement of the pattern from its undisturbed position. Since a wave is a dynamic concept (a wave is never stationary), its description is not complete without an indication of how fast the energy moves. The *speed* of a wave is the distance that the disturbance travels divided by the time elapsed. Since one wavelength passes a point in one period, the speed of the wave is equivalent to the wavelength divided by the period. Another expression for speed is wavelength times frequency.

The wave observables that can be measured are summarized in Table 6–1.

TABLE 6–I MEASUREMENTS OF WAVES

QUANTITY	SYMBOL	METRIC UNIT		
Wavelength	L	meters		
Frequency	f	waves/second,	cycles/second,	hertz
Period	T	seconds		
Amplitude	A	meters		
Speed	v	meters/second		

And the basic relationships among them are:

$$f = 1/T \quad \text{or} \quad T = 1/f$$
$$v = L/T \quad \text{or} \quad v = fL.$$

It is clear from the Slinky demonstrations that a single wave pulse is produced for each to-and-fro motion of the hand. Repeated to-and-fro motions of the wave source result in a succession of wave pulses. Such a train of waves is periodic only if the source is oscillating in a continuous, regular manner. Although few naturally occurring waves are strictly periodic, it is convenient for us in studying waves to assume that we are dealing with a continuous, infinite train of periodic waves and to represent the wave phenomena generally with a static wave form (Figure 6–4).

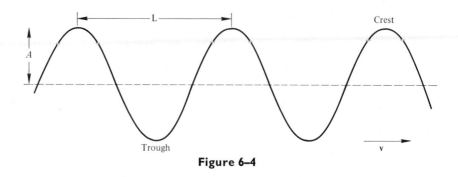

Figure 6–4

The fact that Figure 6–4 looks similar to the graphs of the sine and cosine functions in trigonometry is not accidental; the sine and cosine functions are periodic.

WATER WAVES

The waves with which you are most familiar from everyday experience are water waves. Water waves are produced naturally in a large body of water by the wind and to a lesser extent by the gravitational pull of the sun and moon. Water waves can also be produced for closer investigation in the laboratory by disturbing water in a tank. The tank, called a ripple tank, is usually rectangular in form with a glass bottom. If a ripple tank is available, you should use the water waves experiment as a guide and carefully study the waves generated.* If you do not investigate water waves in detail, then you should recall from experience what you have observed about natural water waves.

Your observation should lead you to conclude that the study of water waves is not very exciting as long as the depth of the water stays about the same and the waves encounter no obstacles. In the absence of discontinuities in the water, there is no change in the waves' wavelength, speed, direction, and so forth. It is the encounter of waves with discontinuities (and with each other) that makes the study of waves interesting. Think of some wave encounters you have seen and what the results of the encounters

---

* It is recommended that Experiment 10 in the lab manual be done at this time.

have been. Think of waves encountering shallower water (as they approach a beach), a small boat sitting in the middle of a lake, or a solid, impenetrable wall. In these situations, and others you can think of, something happens to the wave in the sense that some of its properties are modified by the encounter. These interactions of waves with discontinuities and with each other therefore merit closer examination.

WAVE INTERACTIONS

Generalizing our experience with water waves, we can conclude that as long as a wave moves in empty space or even in a homogeneous medium, there is no change in its wavelength, frequency, speed, or direction of travel. However, when a wave encounters some change in the medium through which it is moving, it does itself change. These wave changes can be summarized as follows:

(1) *Refraction,* the change in direction of a wave as it crosses a boundary.
(2) *Reflection,* the turning back of a wave at a boundary that at least some of the energy cannot cross.
(3) *Diffraction,* the bending of a wave into the geometrical shadow of an obstacle.
(4) *Interference,* the combined effect of two or more waves passing through the same space at the same time.

Refraction. To understand what happens in the refraction interaction, let us push a board with constant force across a table top from a smooth region into a rougher area. If the board encounters the rough area perpendicular to the direction it is moving, the entire board slows, and it keeps moving in the same direction at a slower speed (Figure 6–5).

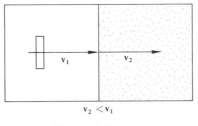

$v_2 < v_1$

Figure 6–5

If the board is pushed with constant force but encounters the rough area obliquely (Figure 6–6), so that the lower end of the board encounters the rough surface first, the lower end slows while the remainder of the board continues to move at its original speed. The top of the board begins to swing around to the right, and continues to do so until the entire board is on the rough surface, at which time all parts are again at the same, slower speed. (The sliding board model is not perfect, for the wave front "kinks"

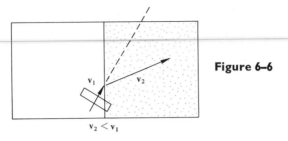

Figure 6–6

$v_2 < v_1$

as it crosses a discontinuity, as you have observed in the case of water waves.) In this encounter, the direction in which the board was moving has been changed. This change in direction is called refraction. It is obvious that a greater change in speed will result in a greater change in direction.

You should be able to reverse the direction of motion of the board and thus analyze what happens when it moves from the rough to the smooth surface.

Reflection. Consider rolling marbles at a flat metal plate. First let the direction of motion of a marble be perpendicular to the plate. In this case, after its encounter with the plate, the marble retraces its original (incident) path, and this we call reflection (Figure 6–7).

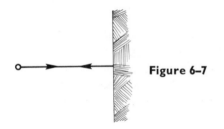

Figure 6–7

If we now let the marble be incident on the plate at an oblique angle, we observe that it comes off the plate at an oblique angle away from the direction from which it approached the plate. It is practical to measure these angles from the normal (perpendicular to the surface). If we change the angle of incidence, the angle of reflection from the plate also changes in an apparently regular fashion (Figure 6–8). On closer examination, we

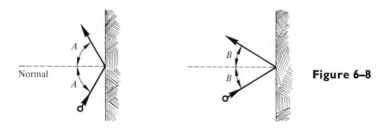

Figure 6–8

find that the angle at which the marble is reflected is always equal to the angle at which the marble is incident.

At this point, you might like to pause and consider our development of refraction and reflection. We began by stating that these are wave interactions, yet we have demonstrated them using particles. It is true that both particles and waves experience refraction and reflection. Here the similarities end, however. There is no particle analogy to either diffraction or interference.

Diffraction.   The diffraction interaction of waves with a discontinuity is most clearly demonstrated if you recall that when water waves encounter a barrier they bend into the shadow region of the barrier (Figure 6–9). This bending into the geometrical shadow is called diffraction.

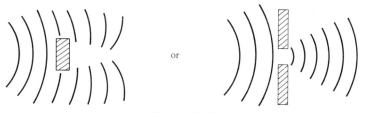

Figure 6–9

Since this interaction cannot be demonstrated in any simpler fashion, we will have to be satisfied at present with the understanding that this is simply a phenomenon that is an inherent property of waves.

Interference.   The interactions we have discussed thus far have all been waves encountering matter. Interference is a wave-wave encounter. It occurs when waves traveling in different directions arrive at the same point in space at the same time. The outcome of this encounter is based on the fact that the propagation of a wave is in no way affected by the presence of other waves in the same space. Thus, two waves reaching the same spot similtaneously pass unaltered through the point. However, the disturbance at the point where the two waves cross is the combination of the disturbance produced by each wave alone. For example, if a crest of one wave arrives at the same time as the crest of another wave, the result is a bigger crest. Likewise, two troughs arriving together produce a deeper trough. This increased disturbance is called *constructive interference*. On the other hand, if a crest and trough arrive simultaneously at a point, they tend to cancel each other, and therefore result in a reduced disturbance. This is known as *destructive interference*.

Consider the case of two continuous wave trains of the same wavelength being produced by two sources. If these two waves move through the same space, interference will occur. However, there is an important variable to be considered in the analysis, and that is the time relationship between the two waves. The time factor is referred to as the *phase*. When two sources or two waves are producing the same kind of disturbance at the

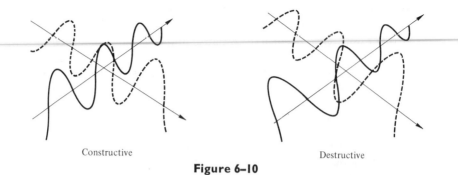

Constructive   Destructive

Figure 6–10

same time, they are said to be *in phase* with each other. Therefore, if two waves of the same wavelength maintain a constant phase relationship to each other (for example, crests always leave the two sources simultaneously), then constructive interference will always occur at the same points in space and destructive interference will always occur at other fixed points. In this case, a stationary interference pattern will be established. Two sources that have the same frequency (produce waves of the same wavelength) and that maintain a constant phase relationship to each other are said to be *coherent* sources. If two waves having the same wavelength and same phase relationship also have the same amplitude, then at some points the crests of one will exactly nullify the troughs of the other, and there is no disturbance at these points. These points of complete destructive interference are called *nodes*.

Two sources that have the same frequency but do not maintain a constant phase relationship produce waves that interfere but that do not establish a stationary interference pattern. Such incoherent sources produce waves for which the crest-crest relationship (constructive interference) at a given point becomes a crest-trough relationship (destructive interference) at the same point at a later time. Therefore, a stationary interference pattern is not established.

Note that whereas the waves are in step when constructive interference occurs, they are half a wavelength out of step when destructive interference takes place. How do two coherent waves become out of step? The crest-trough relationship of two waves that were originally in phase depends on the distance each has traveled relative to the other. If they have traveled the same distance, or one has traveled an integral number of wavelengths more than the other, then crests are together or troughs are together. However, if one wave has traveled one-half wavelength farther than the other, then the crest of one will superpose on the trough of the other and will therefore result in a point of no disturbance.

Is interference anything more than just an interesting phenomenon? Does it add anything to our understanding of physical reality? Suppose that we encounter some strange, new form of energy. One of the things we would like to know about this phenomenon is whether it propagates energy by means of particles or waves. The experiment that will decide this beyond any doubt is one involving interference. We know of no way in which two particles can cancel each other as waves do in destructive interference. Therefore, when a given phenomenon exhibits interference, we know that its energy is waves rather than particles.

How is interference experienced? What does an interference pattern look like? Consider a ripple tank in which there are two sources, S_1 and S_2, of water waves that are coherent and of equal amplitude. Figure 6–11 is an overhead view of the two identical wave patterns produced. The solid lines represent crests and the dashed lines represent troughs.

Figure 6–11

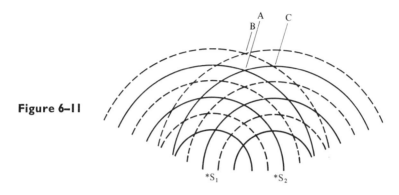

Direct your attention to point A. Here a crest from S_1 arrives simultaneously with a crest from S_2 and the water is disturbed strongly. At point B, two troughs arrive simultaneously and the water is also disturbed strongly. But what happens at point C? Here a crest from S_2 arrives at the same time as a trough from S_1, and therefore the water is not disturbed. Sitting in a boat at C, you would feel no rocking due to waves. If you will now analyze the diagram for all points at which the water is not disturbed, you will find the pattern shown in Figure 6–12. This pattern is readily observed in a ripple tank using two coherent wave sources or using one wave source and a barrier with two openings.

Figure 6–12

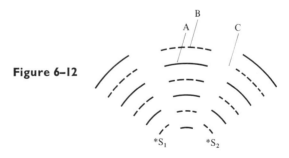

EXERCISES

1. The period of rotation of the earth on its axis is 24 hours and of revolution in its orbit is 365 days. What is the frequency of each motion?

2. From a study of waves on a Slinky, it is found that 20 waves pass a particular point in 10 seconds. It is estimated that the distance from one crest to the next is 0.3 meter. (a) What is the frequency of these waves? (b) What is the period? (c) What is the wavelength? (d) What is the speed?

3. From an airplane, swells (crests) of waves in the ocean are estimated to be 200 meters apart. It is found that a swell passes a given point every 10 seconds. (a) What is the wavelength of these waves? (b) What is the period? (c) What is the frequency? (d) What is the speed?

4. Standing on a dock, you observe water waves passing underneath. You count 30 wave crests passing directly below in one minute and judge the crests to be approximately 0.5 meter apart. (a) What is the frequency of these waves? (b) What is the period? (c) What is the speed?

5. You recall that the motion of particles is described by $d = vt$ and that this motion can be analyzed graphically by plotting d as a function of t. Then the speed is the slope of the curve. For waves, the corresponding relationship is $L = vT$. Determine graphically the speed of the waves represented by the following data:

FREQUENCY (Hz)	WAVELENGTH (m)
10	2.1
20	0.90
30	0.68
40	0.50
50	0.39

What is the wavelength of a wave of frequency 25 Hz?

6. Consider a wave moving in a medium at a speed v_1. Draw a diagram showing how the refraction takes place when the wave passes into a medium in which its speed *increases* to v_2.

7. A wave approaches a barrier at an angle of incidence (measured from the normal) of 60°. Draw a diagram to scale showing the direction of the reflected wave.

8. Explain the following situation in terms of wave interactions. You are sitting in a boat on a lake. There is a solid barrier between your boat and motor boats going up and down the lake. The waves produced by these boats splash against the barrier but do not pass through it, yet your boat, in the shadow of the barrier, is rocked each time a boat passes.

9. The interference pattern of two water waves of the same frequency and amplitude is an array of lanes of disturbed water alternating with lanes of undisturbed water and radiating from the sources. With the aid of a sketch in which you represent the crests and troughs of each wave, show how the lanes of undisturbed water are formed. Since the undisturbed water has no kinetic or potential energy, what has become of the energy of the wave in that region?

7

THE WAVES WE HEAR

Think for a moment about how waves are produced. The waves we studied with the Slinky and the ripple tank were produced by a to-and-fro motion. Further study reveals that the source of any wave is some kind of such vibratory motion.

PRESSURE WAVES

Suppose we do work that results in the deformation of the object we work upon. If the deformed material is elastic, then it will return to its original configuration after the deforming force is removed. Thus, work done on the object results in the object's acquiring energy in the form of vibratory motion. Such a vibrating object might be a stretched string, an air column, or a stretched membrane. If the object is in a medium such as air, then as it vibrates it exerts pressure on the particles nearby. These particles in turn exert pressure on the particles adjacent to them, and so on. Thus, the energy of the vibrating object is transmitted through the surrounding medium by a series of pressure condensations and rarefactions. When these pressure variations encounter an obstacle in the medium, they are capable of giving energy to it and causing it to vibrate in a manner similar to that of the source that produced the pressure variations. Energy has thus been propagated from one point to another with no net transfer of matter. Each air molecule experiences an ordered vibration superposed on its random thermal agitation and passes this ordered disturbance along to the next molecule. These moving pressure variations are therefore longitudinal waves. An understanding of this type of wave is important because sound is a longitudinal pressure wave.

WHAT IS SOUND?

The term "sound" means different things to different scientists. To the physiologist or psychologist, sound is the sensation produced when a pressure disturbance of proper frequency reaches the ear; to the physicist sound denotes the disturbances (waves) themselves, rather than the sensations produced.

Not all pressure variations that reach an ear are capable of stimulating the sensation of hearing. The human ear responds only to disturbances that have a frequency between about 20 hertz and 20,000 hertz. Pressure waves of frequencies below these audible waves are called infrasonic; those above the audible frequencies are called ultrasonic.

To summarize the production, transmission, and reception of sound, let's analyze the sound from a tuning fork across the room. A tuning fork is made to vibrate with a given frequency (say, 500 hertz) when it is struck lightly. In hitting the fork, we do work on it and therefore give energy to it in the form of vibratory motion. As it vibrates at 500 Hz, it collides with air molecules adjacent to it and causes them to vibrate at 500 Hz. These molecules in turn collide with the molecules next to them, thus passing the disturbance along to them. This 500 Hz disturbance is passed from molecule to molecule until the molecules that were originally adjacent to your eardrums are vibrating at 500 Hz. These air molecules exert pressure on your eardrum and cause it to vibrate at the same frequency, and this vibration is transferred through a delicate bone structure, a liquid, and nerves to produce the sensation of hearing in your brain.

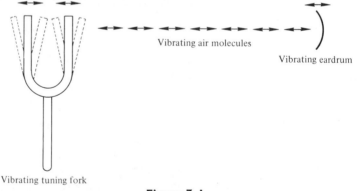

Vibrating air molecules

Vibrating eardrum

Vibrating tuning fork

Figure 7–1

Another mechanism for receiving sound waves is the microphone. In the microphone, a diaphragm is set into vibration, and these vibrations produce electrical pulses. The pulses are actually in the form of voltage variations with the same frequency as that of the diaphragm and thus of the air molecules. These electrical variations are then usually made larger by an amplifier and sent to the speaker, where they cause another, larger diaphragm to vibrate at the same frequency as the microphone but with greater amplitude, thus producing a sound of greater intensity.

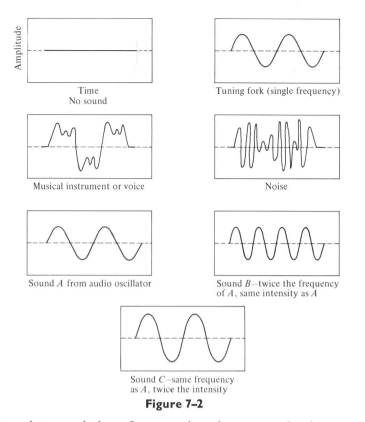

Figure 7-2

Time
No sound

Tuning fork (single frequency)

Musical instrument or voice

Noise

Sound *A* from audio oscillator

Sound *B*—twice the frequency
of *A*, same intensity as *A*

Sound *C*—same frequency
as *A*, twice the intensity

The voltage variations from a microphone can also be sent to an oscilloscope and used to deflect a beam of electrons. When these electrons strike the fluorescent screen of the oscilloscope, they cause light to be emitted. In this way, we can observe the graph of the amplitude of the sound variations as a function of time.

It will be instructive at this point to observe the graphs on an oscilloscope screen of sounds produced by tuning forks, musical instruments, voices, noises, and an audio oscillator (an electronic device capable of producing sounds over a wide range of frequencies and intensities). As you study these graphs (Figure 7–2), look for significant differences in the patterns.

An interesting experiment using an audio oscillator in conjunction with a speaker is to investigate the response of your ear to various frequencies. You will discover that there are frequencies of pressure variations below which and above which your ear does not respond. In addition, you will find that your ear is more sensitive to some frequencies than to others. A "normal" ear is most sensitive to frequencies somewhere around 3000 Hz.

Sounds that have different frequencies are said to differ in *pitch*, and there is very nearly a one-to-one correspondence between pitch and frequency. If one sound has twice the frequency of another, then the ear will interpret it as having twice the pitch. A sound with twice the pitch of another is said to be one octave higher than the other.

Another characteristic of sound to which the ear is responsive is *loudness*. The loudness of sound depends on both its intensity* and its frequency. Although for a given frequency the loudness increases with increasing intensity, the relation between loudness and intensity is not simple. Because loudness depends on the ear and the judgment of the listener, it cannot be measured objectively. We therefore specify the magnitude of sounds by intensity levels. Since the ear is sensitive to an enormous range of sound intensities, it is convenient to define a unit of intensity level of sound in terms of the logarithm of intensity. This unit is called a *decibel* (dB) and is defined by $L = 10 \log I/I_0$ decibels, where L is the intensity level, I is the intensity being considered, and I_0 is a standard minimum intensity.

There is an amplitude below which the auditory nerves are not stimulated. This lowest intensity (I_0) to which the ear responds is called the "threshold of hearing" and has a value on the standard intensity level scale of 0 decibels. At the other extreme of intensity is the "threshold of pain" (120 dB), which is the highest intensity sound the ear can receive without being damaged. The ratio of the intensity of the loudest and the softest sound to which the ear responds is 10^{12} to 1! The amplitude of the sound waves at the threshold of hearing is less than the diameter of an atom. What a sensitive instrument the human ear is!

The approximate intensity levels of some common sounds are shown in Table 7–1.

TABLE 7–I INTENSITY LEVELS

SOUND	INTENSITY LEVEL (dB)
Hearing threshold	0
Whisper	20
Quiet radio in home	40
Conversation at few feet	60
Loud shouting at few feet	80
Electronic "rock" group at several feet	100
Pain threshold, say a few hundred feet behind a jet plane at take-off	120

The relationship between intensity level and frequency varies from person to person; however, for an "average" ear the relationship is approximately as shown in Figure 7–3. Note that the frequency scale is logarithmic.

The loudness difference between two sounds of the same frequency can be specified in decibels by the relation: loudness difference in decibels = $10 \log I_1/I_2$.

---

* Intensity is the power transmitted through a unit area normal to the direction of propagation. Intensity is proportional to the square of the amplitude.

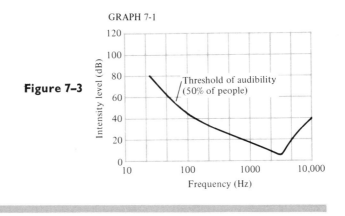

GRAPH 7-1

Figure 7–3

Threshold of audibility
(50% of people)

Intensity level (dB)

Frequency (Hz)

Example: What is the loudness difference between two 500 Hz tones, one of which is 1000 times the intensity of the other?

Answer: Loudness difference $= 10(\log 1000/1)$ dB $=$ $10(3)$ dB $= 30$ dB.

Each tenfold increase in the intensity (power per unit area) adds 10 decibels to the loudness level of the sound. Thus, if the level of sound A is 20 dB higher than sound B, then sound A involves 100 times as much energy per unit area per second as does sound B.

Can sound be described completely in terms of pitch and loudness? Suppose that a note of a given frequency is produced on both a guitar and a trumpet and that the two sounds have the same intensity. Even an untrained ear has no difficulty distinguishing between the two instruments. The characteristic of sound that allows us to distinguish between sounds from different sources is *quality* (or timbre). Difference in quality arises from the fact that most sounds are not simple, but are complex combinations of many simple waveforms. Look again at Figure 7–2 and note the differences in quality.

SOURCES OF SOUND

Any finite collection of particles, whether they constitute a solid, a liquid, or a gas, has certain normal modes of oscillation. This means simply that there are only certain restricted ways in which the body can vibrate. These possible modes are determined by such things as shape, elastic properties, and boundary conditions of the body.

As an example, consider the vibration of a stretched string. Normally, in order for a string to vibrate, it must be fixed at both ends. This means that the two ends of the string must be points of no vibration. Such points of zero displacement in a vibrating object are called *nodes*. The points of

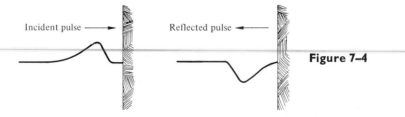

Figure 7-4

maximal displacement (amplitude) in such an object are referred to as *antinodes*. Another restriction that the boundary condition of fixed ends places on the vibration of the string is a result of the reflection of the disturbance at the ends. A transverse wave moving along the string encounters a barrier that it cannot penetrate at the ends. It is therefore reflected, and sends a wave back in the direction from which it came. Furthermore, although the reflected wave has the same frequency as the incident wave, it is exactly out of phase with it (Figure 7-4). Thus, a crest on the incident wave becomes a trough on the reflected wave.

We can have two waves (incident and reflected) of the same frequency moving through the same medium at the same time. As we have seen previously, the two waves interact to create interference. However, this is a special case of interference. Two waves of the same frequency and amplitude moving in opposite directions interact to produce a stationary interference pattern, which is called a *standing wave*. It is easy to determine the wavelength of a standing wave in a string, since it is the distance between alternate nodes; that is, the distance from one node to the next node is one-half wavelength (Figure 7-5).

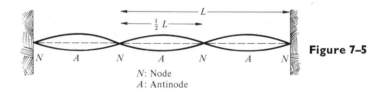

Figure 7-5

N: Node
A: Antinode

The simplest, or fundamental, mode of vibration for a string fixed at both ends is a standing wave with a node at each end and an antinode in the middle. The frequency corresponding to this fundamental mode depends on the density of the string, the length of the string, and the force with which it is stretched (tension). A formula can be derived that shows that the fundamental frequency increases as the tension increases and decreases as the length and density increase. For a string of given density,

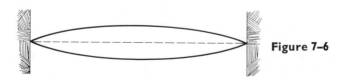

Figure 7-6

length, and tension, no frequency lower than the fundamental can be produced. This fundamental frequency is sometimes called the first harmonic. However, higher modes of oscillation (and therefore higher frequencies) can be produced in this given string by the manner in which it is excited into vibration. The frequencies of the higher modes are referred to as overtones or higher harmonics, and each frequency is an integral multiple of the fundamental frequency.

Figure 7–7 shows the first three possible modes of vibration for a string of length s. L is the wavelength.

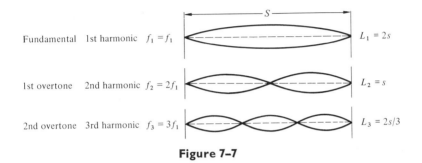

Fundamental   1st harmonic $f_1 = f_1$ $L_1 = 2s$

1st overtone   2nd harmonic $f_2 = 2f_1$ $L_2 = s$

2nd overtone   3rd harmonic $f_3 = 3f_1$ $L_3 = 2s/3$

Figure 7–7

Musically, the notes that a string of a given density can produce depend on the tension, the length of string vibrating, and the manner in which the string is excited.

If we ignore the percussion instruments, most of which do not produce harmonics and thus are not musical in the strictest sense, then the source of sound in a musical instrument is either a vibrating string or a vibrating air column.* An air column also has normal modes of oscillation, but it oscillates longitudinally instead of transversely. However, this does not affect the analysis. The principal restrictions are the length of the column and the boundary conditions. A vibrating air column can be either closed at one end and open at the other (closed tube) or open at both ends (open tube). An end that is closed is like the fixed end of a string and therefore must be a node; however, an open end must be an antinode. The first three modes of oscillation are shown in Figure 7–8 (N = node; A = antinode).

If f_0 is the fundamental frequency of an open column, then the normal frequencies are f_0, $2f_0$, $3f_0$, and so on. However, for the closed column, the normal frequencies are f_c, $3f_c$, $5f_c$, and so on. Therefore, all harmonics can be excited in an open pipe, whereas only odd harmonics can be excited in a closed pipe.

In a musical instrument, several modes of vibration are usually excited simultaneously. The several resulting waves superpose to produce the characteristic quality of the sound. The quality thus depends on the number and intensities of the harmonics produced.

* Electronic instruments (for example, electronic organs) are exceptions.

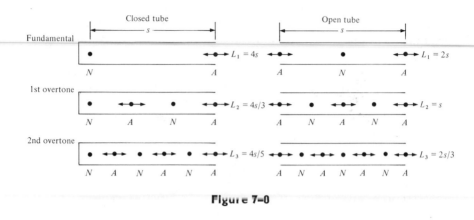

Figure 7-0

RESONANCE

Energy must be supplied to an object to cause it to vibrate. The energy producing the vibrations may occur at regular intervals and thus have a frequency of its own. When the frequency of the energy source is the same as one of the natural frequencies (normal modes of oscillation) of the vibrating body, the body vibrates with maximum amplitude. This condition of maximum energy transfer is resonance. An example of resonance is pushing a child in a swing. You get the swing moving highest and most smoothly when the frequency of your pushes is the same as the natural frequency of the swing. In sound, resonance is involved in using sounding boards to amplify the vibrations produced by vibrating strings in pianos, violins, and guitars. Resonance is particularly important in air columns, since an air column must be set into vibration by some vibrating object, such as a reed or lips. The air column can be caused to resonate with the exciting source by adjustment of the length of the vibrating air column.

Since an air column of variable length has numerous natural modes of vibration, many resonances can be obtained for a given frequency by adjusting the length of the column.

Example: Suppose that we want to make the air in a closed tube vibrate, and we place a vibrating tuning fork near the open end of the tube. The length of the air column can be adjusted by moving a piston that forms the closed end. If the tuning fork has a natural frequency of 400 hertz, how long should the air column be in order for it to resonate (receive maximum energy from the tuning fork)?

Answer: The length in air of the sound wave produced by the tuning fork is $L = v/f = 340$ m/sec $\div$ 400 Hz =

0.85 meters. (We have assumed a speed of sound in air of 340 m/sec.) The fundamental mode of oscillation of a closed air column occurs for a length $s = L/4$. Therefore, the resonant length of the air column is $s = 0.85$ m$/4 = 0.21$ m. When the air column is adjusted to a length of 0.21 meter, a standing wave is established in the air column and it produces its own sound. Resonance will also occur for air column lengths of $3L/4 = 0.64$ m, $5L/4 = 1.06$ m, and so on.

THE SPEED OF SOUND

The speed with which the energy is propagated is an important characteristic of any wave phenomenon. How fast does sound travel? Everyday experience tells you that it moves very fast, but you also know that sound travels much more slowly than light (you see the lightning flash before you hear the sound of thunder).

There are various methods of measuring the speed of sound. One of the simplest methods is to measure the wavelength of a sound wave of a given frequency and then use $v = fL$ to calculate the speed.* Speed is constant for a given medium and temperature. Thus, the speed of sound does not depend on frequency or intensity, but it is a function of the temperature and nature of the medium through which it is passing. The speed in air at 0°C is about 330 m/sec; at 20°C it is approximately 340 m/sec ($\sim$750 mi/hr or $\sim$1100 f/sec). The speed of sound in water is nearly five times as great as in air, and in iron it travels fifteen times as fast as it does in air.

THE WAVELENGTH OF SOUND

The frequency and wavelength of any wave are related through the equation $v = fL$. Since the speed of sound is constant for a given medium (usually air) and does not change greatly with small temperature variations, this formula allows us to determine the wavelength corresponding to a given frequency. Therefore, the wavelength of the lowest audible frequency sound wave is approximately $L = 340$ m/sec $\div 20$ Hz $= 17$ m. The wavelength of the highest audible frequency sound wave is approximately $L = 340$ m/sec $\div 20,000$ Hz $= 0.017$ m $= 1.7$ cm.

---

* It is recommended that Experiment 11 in the lab manual be done at this time.

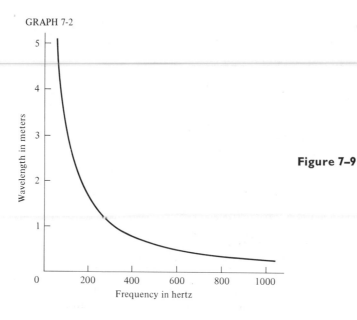

GRAPH 7-2

Figure 7–9

Figure 7–9 shows the relationship between wavelength and frequency (up to a frequency of 1000 hertz) of sound waves, assuming that the speed is a constant value of 340 m/sec.

INTERFERENCE OF SOUND WAVES; BEATS

We considered the interference of sound waves in connection with the phenomenon of standing waves. In that situation, the interacting waves were moving in opposite directions. Interference of sound waves can also occur when the waves are moving in the same direction. If two sources of sound have the same frequency and maintain the same phase relationship to each other (that is, condensations always leave both sources at the same time), then condensations and rarefactions will remain in the same relationship to each other in the wave field. If two condensations or two rarefactions arrive simultaneously at the same point, the amplitude is increased and the resultant sound is louder. If a condensation arrives at a point simultaneously with a rarefaction, then the resultant sound is not as loud as that of either source alone. If the two sound waves have the same amplitude as well as the same frequency and phase relationship, then a condensation of one wave interacting with a rarefaction of the other results in complete cancellation. Under these conditions, there will be points of no sound in the field around the sound sources. This corresponds to the lanes of undisturbed water you see when two sources vibrate the water in the ripple tank. This interference of sound can be demonstrated by connecting a sound source to two identical speakers placed about 1 meter apart. You can hear the zones of constructive and destructive interference if you stand a few meters in front of the speakers and move your head slowly.

An interesting manifestation of interference of sound occurs when two sounds of slightly different frequencies pass through the same space simultaneously. Under these conditions, a stationary interference pattern will not be established; rather, the two sounds will interfere and produce pulsations with a frequency equal to the difference between the frequencies of the two interfering waves. These pulsations are called *beats*.

In order to understand the phenomenon of beats, consider what happens when a sound X of, say, 200 hertz and a sound Y of 220 hertz reach the ear simultaneously. In one second, X will have 200 condensations and rarefactions. In the same second, Y will have 220 condensations and rarefactions. Therefore, 20 times per second condensations (and rarefactions) of X will occur with condensations (and rarefactions) of Y, and the resultant will be increased amplitude (intensity). Likewise, 20 times per second condensations (and rarefactions) of X will occur with rarefactions (and condensations) of Y. The resultant will be reduced intensity. Therefore, the intensity of the 200 hertz and 220 hertz sounds increases and decreases with a frequency of 20 hertz. This 20 hertz variation is the beat frequency that can be heard and also displayed on an oscilloscope.

What about the other interactions of sound waves—reflection, refraction, and diffraction? Probably the clearest demonstration of the reflection of sound waves is the "echo" phenomenon. However, the reflection interaction is of much greater importance to acoustical engineers who must design auditoriums and concert halls to reduce the standing waves that result from reflections of sound from walls, ceiling, and floor. You know that draperies, carpets, and even people reduce the reflection of sound in an ordinary room. You have also experienced the diffraction of sound waves, probably without being aware of it, when you have heard sound around a corner in the geometrical shadow of the sound source. We are less familiar with the refraction of sound. Refraction occurs, for example, when sound waves slow in speed when passing from warm air to cooler air.

THE DOPPLER EFFECT

One of the most interesting phenomena associated with waves occurs when either the source of the waves or the observer of the waves is moving relative to the other. This effect of relative motion between source and observer is called the *Doppler effect* after the man who first analyzed the phenomenon. You have experienced the Doppler effect in the sudden drop in pitch of a siren as it passes you (Figure 7–10). When the source is at rest relative to the observer, the observer receives the true pitch of the source. However, when the source is moving toward the observer, the waves are crowded together, and the observer receives more disturbances per time interval and thus hears a higher frequency. By similar argument, when the source is receding from the observer, it is moving away from the disturbances as they are being produced, and therefore the observer receives fewer vibrations per unit time. The difference between the true

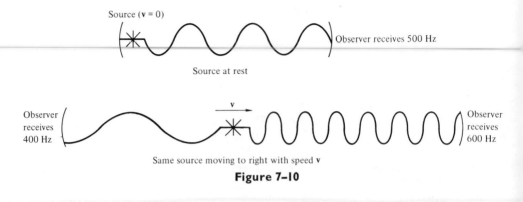

Source (v = 0)

Observer receives 500 Hz

Source at rest

Observer receives 400 Hz

v

Observer receives 600 Hz

Same source moving to right with speed v

Figure 7–10

pitch and the pitch heard by the observer depends on the relative speed between source and observer. The Doppler effect is therefore also experienced by an observer who is approaching or receding from a stationary source.

EXERCISES

Assume that the speed of sound is 340 m/sec.

1. You must have heard the old question about whether or not there is a sound when a tree falls in the forest where there is no one to hear it. Well, what about it? Is there a sound?

2. Sketch the pattern for a sound that has one half the frequency and twice the intensity of the sound shown on the oscilloscope screen in Figure 7–11.

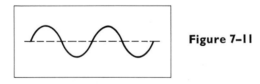

Figure 7–11

3. What is the approximate difference in intensity level in decibels at a cocktail party and at a "rock dance" (near the electronic bandstand)?

4. What is the loudness difference in decibels between two 1000 Hz sounds, one of which is 100 times the intensity of the other?

5. The third overtone is excited in a guitar string that is 1 meter long. The frequency produced is 2400 Hz. (a) What is the wavelength of the standing wave in the string? (b) What is the lowest frequency this string can produce? (c) How many octaves are there between the fundamental pitch and third overtone pitch?

6. For a vibrating string, a general relation for the frequency of the nth harmonic is $f = nf_1$, where f_1 is the fundamental frequency. Write a general expression

for the wavelength of the standing wave in the string for the nth harmonic for a string of length s.

7. A closed air column is 1.2 meters long. What is the wavelength associated with the fundamental frequency of this tube?

8. An open air column is 1.2 meters long. What is the wavelength associated with the fundamental frequency of this tube?

9. The fundamental frequency of an open air column is 400 Hz. What is the fundamental frequency of a closed air column of the same length as the open column?

10. The following data are from an experiment designed to determine the speed of sound in air by measuring the wavelengths of sound waves of various frequencies. On coordinate paper, plot the wavelength as a function of period and determine from the resulting curve the speed of these sound waves.

FREQUENCY (Hz)	WAVELENGTH (m)
100	3.5
200	1.7
300	1.2
400	0.86
500	0.71

11. You have probably heard that you can estimate how far away a thunderstorm is by counting the number of seconds between the time you see the lightning and the time you hear the thunder and allowing 1 mile for every 5 seconds of elapsed time. Explain why this is true. (The speed of light is so great that you can assume there is no delay in its arrival.)

12. You observe rocks crumbling from a dynamite blast in the side of a mountain and 6.5 seconds later you hear the report of the blast. How far away was the explosion?

13. A short beep is sounded on the horn of a car that is facing a canyon wall. Four seconds later, the reflected wave (echo) is heard. How far is the car from the wall?

14. The ear is able to distinguish two sounds as separate only if they reach it at least 0.1 seconds apart; otherwise they blend into a single sound. What is the minimum distance from a reflecting surface at which an echo can be heard?

15. Concert pitch is 440 Hz (A above middle C). What is the wavelength of this sound wave in air?

16. The frequency of middle C is 256 Hz and that of C above middle C is 512 Hz. Compare the wavelengths of these two sound waves, which differ in pitch by one octave.

17. Why do you not hear interference when sound is coming from the two speakers of a stereo set?

18. A tuning fork of unknown frequency is sounded simultaneously with a fork whose frequency is 256 Hz. A beat frequency of 50 Hz is detected. What is the frequency of the unknown fork?

19. How could the phenomenon of beats be applied to the tuning of a string on a guitar to the frequency of a particular note on a piano?

20. Explain why the Doppler effect occurs when there are a stationary source and a moving observer.

21. What evidence can you give from your pre-physics experience that sounds of different frequencies travel with the same speed?

8

THE WAVES WE SEE
(AND DO NOT SEE)

What is light? Since we are trying to explain and thus understand physical reality, an answer such as, "Light is what we see with," will not suffice. One way of approaching the question is to list every fact you know about light. Try it. Now, from your list, can you say what light is? Chances are, most of the items on your list tell what light does rather than what it is. An important consideration regarding light is the fact that this form of energy reaches the planet Earth from the far reaches of the Universe, whereas sound does not. Since we are convinced that much of the Universe is empty space, a significant problem is to explain how light travels in the absence of matter.

ELECTROMAGNETIC WAVES

One definition of sound is that it is that group of pressure disturbances to which the ear is sensitive. In similar fashion, light can be defined as that group of disturbances to which the eye responds. But what kind of disturbances are these? The general class of disturbances of which light is a member is electromagnetic waves. Whereas sound is a pressure disturbance, light is an electromagnetic disturbance.

The word "electromagnetic" is derived from the words "electric" and "magnetic," which implies a combination of electric and magnetic effects. You will recall that a charge produces electrical effects (fields) in the space around the charge. Furthermore, if the charge is moving, the space around the charge is further altered, and this additional effect on space is a magnetic field. Now suppose that the charge is oscillating, say in the form of an electron in an antenna. This oscillating charge will result in an

191

oscillating electric field and an oscillating magnetic field. It can be shown theoretically and verified experimentally that these electric and magnetic fields combine to produce a disturbance that is propagated through space. This disturbance is an electromagnetic wave. This wave is self-propagating in the sense that a changing electric field induces a changing magnetic field, which in turn induces a new changing electric field, and so on. Thus there is a constant energy interchange between the electric and magnetic fields. When this electromagnetic wave encounters charges in matter, the electric and magnetic fields cause the charges to oscillate in the same manner as that in which the originating charges were oscillating. Therefore, energy has been transferred from one point to another with no net transfer of matter.

There are many levels in the structure of matter at which charges are found—free electrons in a conductor, electrons in an atom, and charged particles within the nucleus of an atom, to mention a few. The speed with which charges oscillate also varies over a wide range. These two factors, the source of charges and the rate of oscillation of the charges, result in different types of electromagnetic waves with different frequencies (or wavelengths). However, all of these various types of electromagnetic waves have certain important properties in common. All electromagnetic waves (1) are produced by moving charges, (2) are transverse waves, (3) do not require a medium for transmission, and (4) travel with the same speed in the absence of matter (vacuum), and this speed is absolute. The speed in free space is indicated with a c and has the value 2.9979×10^8 m/sec, which is 3.00×10^8 m/sec to three significant figures. The absoluteness follows from the fact that no greater speed has been measured and the belief that energy cannot be propagated faster than 3×10^8 m/sec.

Figure 8–1

James Clerk Maxwell (1831–1879)—*Scottish*— *Significant progress toward understanding the physical universe was made when Maxwell synthesized light, electricity, and magnetism during the third quarter of the nineteenth century. He also contributed important ideas to the kinetic theory of gases. Maxwell showed signs of his genius at an early age. He did for electromagnetic phenomena what Newton had done for mechanics. He produced a set of equations that related electricity, magnetism, and light in a single system. Maxwell expanded man's restricted conception of physical reality to include fields as well as material particles. An untimely death from cancer at the age of 48 prevented Maxwell from seeing his theory verified by experiment.*

An arrangement in order of frequencies (or wavelengths) of the various types of electromagnetic waves is known as the electromagnetic spectrum. These electromagnetic disturbances are often called radiation.

Actually, the term "radiation" refers to any spreading out of energy from a source. Thus, sound is radiation. However, the term is generally used only in connection with electromagnetic energy. In the electromagnetic spectrum (Figure 8–2), frequencies and wavelengths are related through the formula $L = c/f$. There are no distinct lines of separation between the various named electromagnetic phenomena, but rather a somewhat fuzzy zone in which one type ends and another begins.

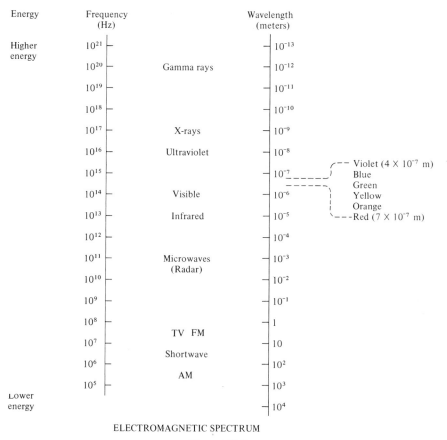

ELECTROMAGNETIC SPECTRUM

Figure 8–2

LIGHT

Now, let's go back to our original question, "What is light?" Light is that group of electromagnetic waves to which the eye responds. The range of these disturbances is summarized as follows:

Frequency: from $\sim 4 \times 10^{14}$ Hz to $\sim 7 \times 10^{14}$ Hz

Wavelength: from $\sim 7 \times 10^{-7}$ meters to $\sim 4 \times 10^{-7}$ meters

Color: from red to violet.

All these waves travel 3×10^8 m/sec in the absence of matter. In a medium, the speeds of all are less than 3×10^8 m/sec, varying with wavelength (color). Red travels fastest in a medium, violet slowest. Thus, in crossing the boundary between two media, red experiences the least refraction, violet the greatest refraction, and all other colors some amount of refraction between that of red and that of violet. This explains why a prism spreads out (disperses) white light into its component colors (Figure 8–3). A rainbow is produced by dispersion and reflection of sunlight by numerous drops of water.

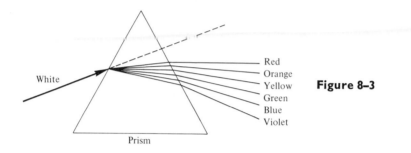

Figure 8–3

The array of electromagnetic energies to which the eye responds thus forms the visible spectrum within the electromagnetic spectrum. Just as different sound frequencies appear to the ear as differences in pitch, so different light frequencies (or wavelengths) appear to the eye as differences in color. The colors that are associated with definite ranges of wavelength are red, orange, yellow, green, blue, and violet. For example, light of wavelength 5.2×10^{-7} meters looks green to the normal eye. Some colors, such as brown, are mixtures of two or more different wavelengths. White is a combination of all wavelengths. Black is the absence of color (and therefore of visible light). The color of a transparent object is determined by the wavelengths it transmits. The color of an opaque object is determined by the wavelengths it reflects; the other wavelengths are absorbed.

GRAPH 8-1

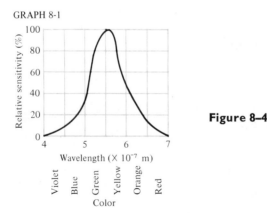

Figure 8–4

For example, a red sweater reflects red and absorbs all other colors incident on it.

Although the eye responds to a range of frequencies from approximately 4×10^{-7} meters to 7×10^{-7} meters, it is not equally sensitive to all frequencies in this range. In fact, the human eye has a maximum sensitivity near the middle of this range at about 5.5×10^{-7} meters (yellow-green). Graph 8–1 shows the response of the "average" human eye to light.

REFRACTION OF LIGHT

In investigating the refraction and reflection of light, we will find it convenient to represent a light wave with a straight line drawn in the direction in which the wave is traveling. Such a line, which is perpendicular to the wave front, is called a *ray*.

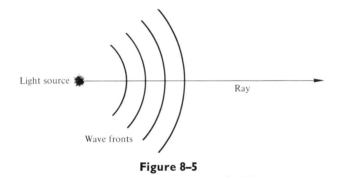

Figure 8–5

As long as light travels in empty space, it undergoes no interactions, and the energy is propagated with constant speed in a straight line. However, when light encounters a discontinuity, at least one interaction, and possibly more, takes place. Interactions will thus occur as light passes from empty space to air. In air, light propagates at a constant but slower speed than in a vacuum. If the energy now crosses the discontinuity from air to glass, another interaction occurs. The light travels at a constant but even slower speed in glass than in air. Therefore, regardless of what other interactions may occur, the light will undergo refraction in these cases, since there is a change in speed. Since a greater change in speed results in a larger change in direction, it is useful to give this change a quantity in terms of an index of refraction, n, which is given by $n = c/v$ where c is the speed of light in empty space and v is the speed in a given medium. Thus, the larger n is, the larger the change in direction at the boundary of the medium. This relation is strictly true only for light of a single frequency (monochromatic light). Although waves of all frequencies (colors) travel with the same speed in empty space, waves of different frequencies travel with different speeds in a transparent medium. Thus, there is an index of refraction for each color.

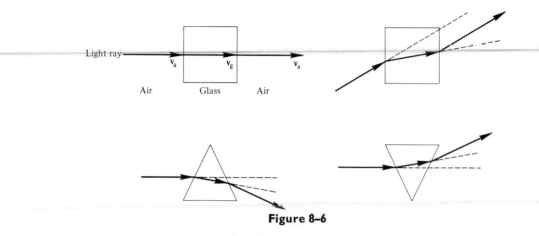

Figure 8–6

Figure 8–6 shows the refraction interaction for monochromatic light crossing air-glass boundaries (hereafter we will assume that all interfaces are air-glass) at various angles. Analyze each situation and observe a demonstration of each if possible. In all cases, v_{glass} is less than v_{air}.

One of the things we would like to do with light is to collect it in a small area. The last two diagrams in Figure 8–6 suggest a method of doing this. Suppose that we set the two prisms base to base and let light rays strike both prisms (Figure 8–7). As you can see, refraction causes the rays to come together.

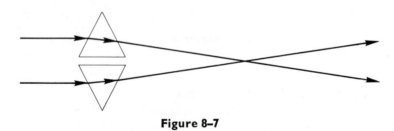

Figure 8–7

If we now combine the two prisms and make the surfaces spherical (cross-sectional arcs of circles rather than straight), we form a converging (convex) lens. This lens is able to focus all parallel rays of one color at a single point. The point at which parallel rays come together is the *focal point* of the lens and the distance from the center of the lens to the focal point is the *focal length*. Parallel rays are obtained in practice by having the source of the rays at a very great distance (ideally, at an infinite distance) from the lens. The results shown in Figure 8–8 can be easily verified with a converging lens and a source of parallel light rays.

An important principle concerning lenses is that the light rays passing through them are reversible. If the source of light in the diagram is at F, the path of each ray would be exactly the reverse of those in the diagram, and therefore would be made parallel by the refraction. From a symmetry argument, it is also clear that the parallel light in the diagram could just as

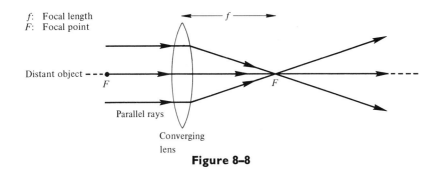

Figure 8–8

well have come from the right, passed through the lens, and come to a focus on the left. Thus, a lens has two focal points. There is one ray that passes through both foci and also passes through the center of the lens. A straight line coinciding with this ray is called the *principal axis*. It is helpful in drawing ray diagrams to include the principal axis as a reference line.

Rays from an object that is relatively near the lens are diverging, not parallel. These rays are focused by the lens, but not at the focal point. There is a simple graphical method for finding out where the rays focus after refraction:

(1) Draw a ray from the object parallel to the principal axis. After passing through the lens, this ray passes through the focal point.
(2) Draw a second ray from the same point on the object, straight through the center of the lens. (Refer to Figure 8–6.)
(3) The intersection point of the two rays is the image point corresponding to the object point.

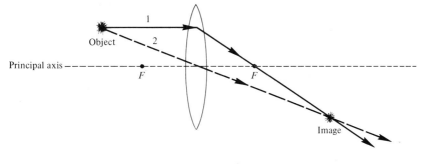

Figure 8–9

Now suppose that instead of a point object we have an extended object. We can find an image point corresponding to every point, and thus we can determine completely the position and characteristics of the image.

From a ray diagram, we can determine: (1) whether an image is erect or inverted, (2) whether it is magnified, diminished, or the same size as the object, and (3) whether it is real or virtual. An image is real if the

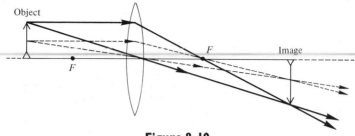

Figure 8–10

light actually passes through the image (can be focused on a screen) and virtual if the light only appears to come from the image (cannot be projected on a screen). One other characteristic, which we will not analyze, is whether or not the image is perverted. Perversion refers to the interchange of left and right, as in the reflection from a plane mirror.

What happens if we place the prisms apex to apex instead of base to base and pass light through this arrangement (Figure 8–11)? The light rays are spread apart instead of brought together. This, therefore, forms a diverging (concave) lens, from which parallel rays appear to diverge from a virtual focus. This can be verified with a demonstration.

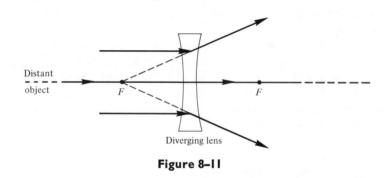

Figure 8–11

The graphical method for finding the position and character of the image formed by a diverging lens is the same as the method that we outlined for a converging lens.

Careful observation should reveal to you that there is some reflection each time light crosses a boundary. It is true that the amount of light transmitted across even a transparent boundary is never 100 per cent.

REFLECTION OF LIGHT

When light waves encounter a boundary across which they cannot all pass, at least some are reflected. If this boundary is in the form of a rough surface, irregular reflection occurs, and this becomes the light by which the surface is seen. Any object, unless it produces its own light, is seen by reflection. If the surface is smooth, regular reflection occurs and a reflected image of the object is produced. Let is now investigate this

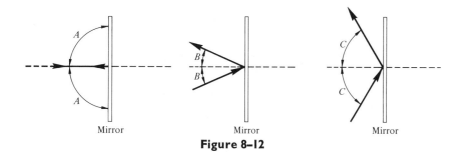

Figure 8–12

regular reflection of light. The reflection from a plane mirror can be easily demonstrated (Figure 8–12).

As we discovered when we rolled a marble onto a plate, the angle of reflection is always equal to the angle of incidence. Furthermore, the reflected ray, the incident ray, and the normal are in the same plane.

Now suppose that we have two parallel rays striking two mirrors that are set at an angle of less than 180° to each other (Figure 8–13). After reflection, these rays will pass through a point. Following the idea we used in connection with lenses, we will now replace the two plane mirrors with a single spherical reflecting surface and thus have a converging (concave) mirror. The analysis of the converging mirror will be similar to that for the converging lens, but we must keep in mind that the light is reflected rather than refracted and that there is only one focal point.

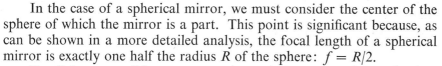

Figure 8–13

In the case of a spherical mirror, we must consider the center of the sphere of which the mirror is a part. This point is significant because, as can be shown in a more detailed analysis, the focal length of a spherical mirror is exactly one half the radius R of the sphere: $f = R/2$.

Parallel rays striking a converging mirror pass, after reflection, through the focal point.

The graphical method for determining the position and nature of the image of a relatively near object formed by a converging mirror is as follows:

(1) Draw a ray from the object parallel to the principal axis. After reflection, this ray passes through the focal point.

(2) Draw a second ray from the same point on the object through the center of the sphere. This ray is reflected back on itself. (Why does this ray return on itself? Hint: See Figure 8–12.)

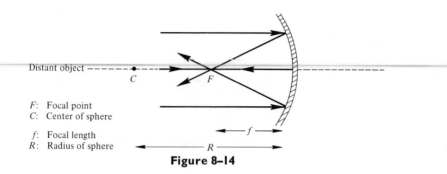

F: Focal point
C: Center of sphere

f: Focal length
R: Radius of sphere

Figure 8-14

(3) The intersection point of the two rays is the image point corresponding to the object point.

This image is real, inverted, and diminished. It is worth noting that these rays, as in the case of the lens, are reversible. If the image is made the object, then its image would be where the object is in Figure 8–15.

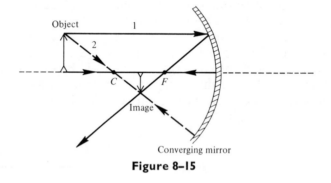

Converging mirror

Figure 8-15

It should be clear to you what will happen to light rays if the outside surface of the sphere is used as a mirror. The results shown in Figure 8–16 can be demonstrated. This diverging (convex) mirror spreads the parallel light rays out the way a concave lens does. The reflected rays appear to diverge from a virtual focal point. The position and character of the image are determined graphically by applying the method that we have outlined for a converging mirror.

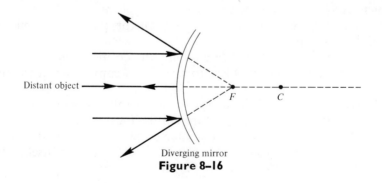

Diverging mirror
Figure 8-16

Optical Instruments.   Both converging and diverging lenses are used in eyeglasses to correct defective vision. A converging lens is used in a camera to focus the image of the object to be photographed on the photographic film. Telescopes have either a converging lens or a converging mirror to gather light and form an image. The image formed by the objective (light-gathering lens or mirror) is then examined with a converging

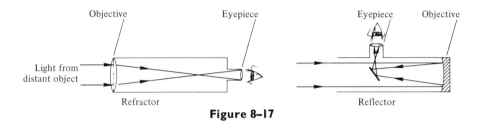

Figure 8–17

lens eyepiece. A telescope with a lens as the objective is called a *refractor* one with a mirror as the objective is called a *reflector*. In a compound microscope, two converging lenses are used to form a greatly magnified image of a small object. The objective forms a magnified image that is magnified a second time by the eyepiece.*

---

* It is recommended that Experiment 12 in the lab manual be done at this time.

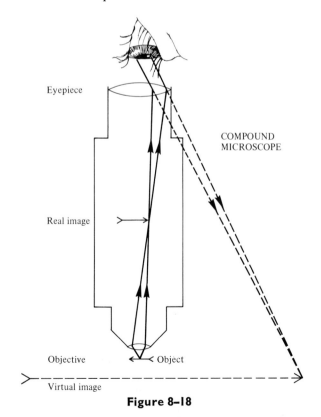

Figure 8–18

DIFFRACTION OF LIGHT

When light waves traveling in a homogeneous medium encounter an impenetrable obstacle, they bend around the edge of the obstacle. Thus, some light appears in the geometrical shadow of the obstacle. This is the diffraction phenomenon, which is most easily visualized with water waves. It is because of diffraction that objects do not cast sharp shadows. The diffraction effect is particularly prominent in the sort of hazy, ragged-edged shadows cast by very small objects. The occurrence of diffraction is most notable when light encounters a very small opening or slit in an otherwise opaque barrier. In this case, as was demonstrated for water waves, the plane of the slit becomes a new source of many new waves, which spread out into the space beyond the opening, and therefore diffraction of light by a slit also produces interference. This phenomenon can be observed if you hold two fingers pressed together close to your eye and look at a bright light source through the opening. It is clear that an analysis of the diffraction of light waves also involves the interference of light waves.

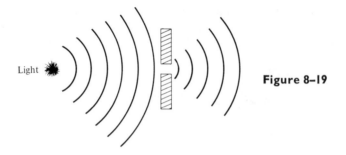

Figure 8–19

INTERFERENCE OF LIGHT

The effects of interference of light are most easily understood by analyzing what happens when monochromatic light from a single source encounters two openings in a barrier (Figure 8–20). Diffraction occurs at each of the slits, and thus there are waves spreading out from each of the slits into the region to the right of the barrier. We are interested in the effects produced by the interference of these waves. We will observe these effects by intercepting the energy on a screen placed parallel to and some distance from the barrier.

Although we cannot observe what is happening between the slits and the screen, we do have a mental picture of the interactions there from our consideration of the interference of water waves. You remember that the result of the interference of water waves was a pattern of lanes of strongly disturbed water radiating from the slits and separated by lanes of undisturbed water. When a lane of increased energy is intercepted by a screen, light energy is reflected by the screen and appears as a bright line on the screen. However, in a lane along which crests and troughs come

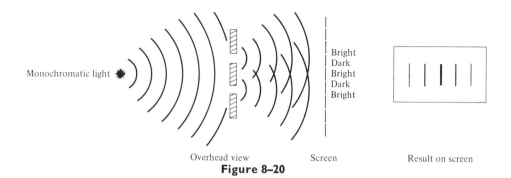

Overhead view   Screen   Result on screen

Figure 8–20

together, there is no resultant energy, and this appears on the screen as a dark line or the absence of light energy. Therefore, the front of the screen will show a series of bright lines separated by dark lines. The center line is brightest, with the intensity of lines decreasing on either side. This is the effect of the interference of light of a single color (wavelength).

Now suppose that we replace the monochromatic light source with a source of two colors, say, red and blue. Since red light and blue light have different wavelengths, crests and troughs will come together at different points for each color. Although there are lanes of maximum energy and minimum energy for each color, these lanes do not coincide. Where these lanes intersect the screen, there will be red lines, blue lines, and dark lines (this can be demonstrated). The argument just developed can be extended to all colors, and this is why the interference interaction separates white light into its component colors.

Increasing the number of diffraction slits in a given space diverges the spectrum and narrows the individual lines. Since both spread of the spectrum and sharpness of lines are desirable features, the number of slits used is usually very large, being on the order of 10,000 openings per centimeter. An arrangement of a large number of parallel, equally-spaced slits is called a diffraction grating.

You will recall that in order to produce a stationary interference pattern, the interfering waves must have the same frequency and maintain

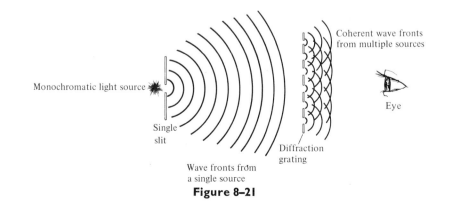

Figure 8–21

a constant phase relationship to each other. In the case of light, where frequencies are on the order of 10^{14} hertz, it is impossible (except for lasers) to have two separate sources in which the waves are coherent. The diffraction grating overcomes this problem by making many sources of waves from a single source of light (Figure 8–21).

Since wave trains that produce interference seem to be identical, what is the cause of the interference interaction? You will remember that the other essential element in the interference of waves is that the waves must travel different distances before they come together. It is in this way that crests and crests meet at one point, crests and troughs at another. This suggests that any mechanism that splits light from a single source into two or more wave sources should produce interference effects. One of the simplest methods of achieving this is to reflect light from the top and bottom of a thin film of transparent material.

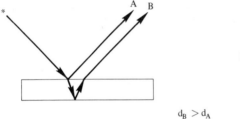

Figure 8–22

$$d_B > d_A$$

In thin film interference, the interfering waves are produced, as shown in Figure 8–22. A source of monochromatic light is split into beam A, reflected from the top surface of the film, and beam B, reflected from the bottom surface. When beams A and B recombine, B has traveled farther than A. The crest-trough relationship for the two beams depends on the thickness of the film. Now consider a thin film that is not of uniform thickness, but has a wedge-shaped cross-section (Figure 8-23). The difference in the distance traveled by the two reflected beams now depends upon the thickness of the wedge at the point at which they are reflected. At one thickness, constructive interference will occur, and at both smaller and greater thicknesses on either side destructive interference will occur. Looking at light reflected from the wedge, we see a pattern of alternating bright and dark lines similar to the diffraction slits pattern.

If we now replace the monochromatic light source with white light,

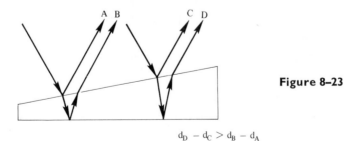

Figure 8–23

$$d_D - d_C > d_B - d_A$$

then whether or not constructive interference occurs and produces a bright line depends not only on the thickness of the film, but also on the wavelength of the light. For example, since red light has a longer wavelength than blue, constructive interference for red occurs at a different film thickness than it does for blue. Thus, a transparent thin film in the shape of a wedge separates white light into its component colors by interference. The colors in soap bubbles and those seen in oil films on puddles in the street after a rain are produced by interference of white light.

POLARIZATION

Once we have established the wave nature of a given phenomenon, how do we determine whether the waves involved are transverse or longitudinal? For waves in a coiled spring, we simply observe the wave pattern in the spring. But what about sound and light? How do we know that sound waves are longitudinal and that light waves are transverse? Since we cannot observe the wave patterns for sound and light, making the decision in these cases is analogous to determining whether waves in a coiled spring are longitudinal or transverse without observing the spring. Think for a moment about how you might determine whether waves in a Slinky are longitudinal or transverse without looking at the spring. You could do it with two bricks, by placing one on either side of the spring. Figure 8–24 shows the view looking down at a Slinky on a table. As you

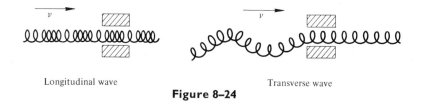

Longitudinal wave   Transverse wave

Figure 8–24

can see, the longitudinal wave is unaffected by the bricks, whereas the transverse vibrations are almost completely stopped. Little energy emerges from the encounter with the bricks when the waves are transverse and in the plane of the paper. Note that vibrations perpendicular to the paper would not be affected by the bricks. Therefore, if transverse vibrations are in random directions, such an encounter with bricks will restrict the vibrations to a plane. An interaction with matter in which transverse waves are restricted to a particular plane of vibration is called *polarization*. To be completely correct, this interaction should be referred to as plane polarization, since there are other types of polarization, such as circular and elliptical. However, when the term polarization is used, it is generally understood to be plane polarization.

Now, we should be able to use this same analysis to establish the transverse nature of light waves, if we could get some bricks of the proper size! Interestingly enough, there are certain natural materials, such as calcite and tourmaline, that have "bricks" of the correct dimensions for light polarization built into their crystalline structure. The most practical polarizing medium for light, however, is a man-made material called Polaroid, which consists of needle-like crystals embedded parallel to each other in plastic. When a beam of light passes through a sheet of Polaroid, its intensity, and thus its energy, is reduced by approximately one half. If this polarized light is now intercepted by a second sheet of Polaroid with its needles perpendicular to the needles in the first sheet, then almost no light emerges from the second sheet. Thus, crossing two polarizing substances essentially stops all the energy in a transverse wave. The energy in a longitudinal wave is barely affected by crossed polarizers.

End-on view of light wave approaching an observer

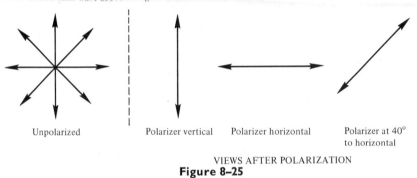

Unpolarized   Polarizer vertical   Polarizer horizontal   Polarizer at 40°
                                                                      to horizontal

VIEWS AFTER POLARIZATION
Figure 8–25

Upon closer investigation, we learn that light is polarized in other types of encounters also. One such encounter is reflection. Light is plane polarized, to an extent depending on the angle of incidence, upon reflection from surfaces such as glass and water. It is for this reason that Polaroid sunglasses are particularly effective in reducing glare from water and snow.

Light is also partially polarized when it is scattered by very small particles in the medium through which it is traveling. Light passing through the earth's atmosphere sets particles there vibrating. These vibrating particles radiate the energy they have received from the light in another direction. This re-radiated light is partially polarized. It is this same scattering process that gives the sky its color. Depending on the size of the particles, a particular wavelength, usually corresponding to blue color, is selectively scattered, and this is the color we see when we look at the sky.

A very interesting effect that certain materials have on polarized light is the rotation of the plane of polarization as the light passes through the material. This property, called optical activity, is exhibited by sugar and by many plastics, among other materials. The amount of rotation of the plane of polarization for a given material depends on the thickness of the material passed through. A striking demonstration of optical activity can be made using various thicknesses of cellophane tape on a piece of glass mounted between two sheets of Polaroid. Different thicknesses of tape produce different colors, and the colors change as the Polaroid sheets are rotated with respect to each other.

THE SPEED OF LIGHT

We stated previously that electromagnetic waves travel in space at a very high, constant speed. Since light is an electromagnetic wave, it follows that in the absence of matter, light travels at this speed of 3×10^8 m/sec. But how do we know? How do we measure a displacement of 300 million meters in an interval of 1 second? Clearly, we must either use very large distances or find a method of measuring very small time intervals. The first reasonably successful determination of the speed of light was done late in the seventeenth century using the large-distance alternative. The distance involved was the diameter of the Earth's orbit around the sun and the time interval was the period between the eclipses of one of Jupiter's moons. In spite of the fact that the diameter of the Earth's orbit was not accurately known, this method gave a value for the speed of light that was within 30 per cent of the currently accepted value.

Most of the later attempts at measuring the speed of light have been directed toward measuring very short time intervals with essentially two types of devices—rotating toothed wheels and rotating multiple-faced mirrors. The time interval is measured in terms of the speed of rotation of the tooth or mirror mechanism required to cause a light beam that is returning from a distant mirror to encounter the next tooth or mirror in the rotating mechanism. Today, somewhat more sophisticated techniques involving such things as molecular spectra and interferometry are employed.

RELATIVITY

Consideration of the speed of light raises an interesting question. With respect to *what* does light travel at a speed of 3×10^8 m/sec? We understand that the speed of a water wave is measured relative to still water or the shore. The speed of sound is 340 m/sec measured relative to air that is not moving. In fact, all our measurements of speed are made relative to something that is assumed to be stationary. To say that the speed of a car is 60 miles per hour implies that the speed is measured relative to the Earth or something attached to it, such as a tree or a building. But of course the Earth itself is moving, rotating on its axis and revolving in orbit around the sun, moving with the sun through the Galaxy, and so on. Does this mean that all motion is measured relative to something that is also moving? Is there some object in the Universe, maybe a star somewhere, that is not moving, so that all motion could be measured relative to it? No such object has been found, and there is reason to believe that such an absolute reference point does not exist. Einstein suggested that there is no hitching post in the Universe. So, relative to what reference point is the speed of light measured?

The idea of relating the speed of light to a fixed medium like sound waves in still air seemed so plausible in the nineteenth century that a medium called "the ether" was postulated and presumed to pervade all space. Experiments were undertaken to detect it. If there is an ether, then the Earth as it revolves around the sun must be moving through this medium. In the 1880's, two physicists devised an experiment to detect the speed of the Earth through the ether by measuring the speed of light parallel to the Earth's motion and perpendicular to it. If a difference existed, they figured, it could be detected by the interference of light waves. This famous Michelson-Morley experiment determined that the speed of the Earth relative to the postulated ether was zero. Therefore, the notion of an all-pervading medium was discarded.

A disturbing consequence of the Michelson-Morley experiment is the conclusion that light travels with the same speed in all directions and that its observed speed is independent of the motion of the observer. Common sense tells us that if an observer who is at rest relative to a light source measures the speed of light as 3.0×10^8 m/sec, then an observer who is approaching the source at 0.1×10^8 m/sec should measure a speed of 3.1×10^8 m/sec. Not so! Both observers measure a speed of 3.0×10^8 m/sec. These experimental results, coupled with new insights into the nature of physical laws, led Albert Einstein to postulate the special theory of relativity in 1905.

The special theory of relativity is a deductive system in which the results of the theory are concluded logically from two simple postulates:

(1) Principles of physics that are valid in one reference system are equally valid in any system that is moving with constant velocity relative to the first system. There is no preferred reference system.

(2) The speed of light (electromagnetic waves, generally) in free space is the same for all observers regardless of their motion.

The first postulate, known as the principle of relativity, implies that in a spacecraft cruising at constant speed in a straight line, the results of any experiment performed will be the same as they would be if the craft were not moving. No experiment performed within a closed system will detect uniform motion of the system. The second postulate is a statement of experimental results.

Figure 8–26

Albert Einstein (1879–1955)—*German-Swiss-American—The year 1905 was a good year for physics. In that year, Einstein published papers on the quantum theory of light and on special relativity, which established a different plateau from which to observe and conjecture. This modest man disdained formality, but enjoyed his pipe, violin, and physics. He gained much fame with his colleagues and the general public. He won the Nobel Prize in physics in 1921, not for the theory of relativity, but for his work on the photoelectric effect. Only time will put into proper perspective the contributions this great genius made to our understanding of physical reality.*

Although these postulates are simple enough, the conclusions that result from them are radical indeed! Our common-sense ideas about absolute space and absolute time are shown to be incorrect. Events that are simultaneous for observer A are not simultaneous for observer B, who is moving relative to observer A. If observer B measures a distance of 1 meter between two points in his system, then observer A, in motion relative to B, will find the distance between the two points to be something less than 1 meter. It is clear, then, that our traditional notion of absolute time and space must be abandoned. Relativity fuses the separate concepts of space and time into the single concept of a space-time continuum. Further development of the deductive system leads us to conclude that the mass of an object is also relative to how fast it is moving. This in turn causes us to reconsider the classic concepts of momentum and energy.

Observers moving with respect to each other will disagree on their measurements of length, time, and mass. What is needed is a method for transforming measurements made by an observer in one system into measurements made by an observer in another system when one system is moving relative to the other. Such transformations can be derived from Einstein's two postulates. The transformation formulas for relativistic length, time interval, and mass are:

$$L_v = L_0 \cdot \sqrt{1 - v^2/c^2} \qquad T_v = T_0/\sqrt{1 - v^2/c^2} \qquad m_v = m_0/\sqrt{1 - v^2/c^2}$$

In each formula, v is the relative speed between the two systems; c is the speed of light; L_0, T_0 and m_0 are values measured by an observer at rest relative to the measured quantity; and L_v, T_v, and m_v are values measured by an observer whose speed relative to the measured quantity is v.

It is not surprising that all these transformation equations involve the ratio of the relative speed between the systems v to the speed of light c. If v is small compared with c, v/c is small and $(v/c)^2$ is even smaller. Thus, for low speeds (everyday kinds of motion), there is little difference between length, time, and mass as measured by stationary and moving observers. As v approaches c, both time interval and mass as measured by the moving observer increase, whereas length decreases.

Length Contraction. Graph 8–2 of $L_v = L_0 \cdot \sqrt{1 - v^2/c^2}$ shows the length of an object as measured by an observer as a function of the speed of the object relative to the observer. The speeds are plotted as fractions of the speed of light c and lengths as multiples of the rest length L_0 (the length as measured by the observer when the object is not moving relative to him). It should be clear from a study of the graph why we are not concerned with length contraction at commonly experienced speeds. On the scale of this graph, the decrease in length with speed is not apparent up to 0.1 c, which is a speed of 3×10^7 m/sec (18,600 miles per second). That is fast enough to travel three fourths of the distance around the Earth at the Equator in 1 second! However, a meter stick moving at a speed of 0.9 c (2.7×10^8 m/sec) would appear to be less than half a meter in length. Length contraction is observed only parallel to the direction of motion. A

GRAPH 8-2

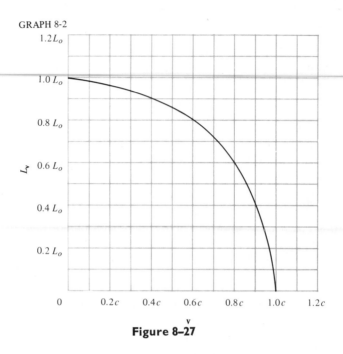

v

Figure 8–27

meter stick that is perpendicular to the direction in which it is moving will not appear to be shorter.

Time Dilation.   The graphical analysis (Graph 8–3) of the variation of time intervals on a clock moving relative to an observer reveals that, rather than being contracted with speed, time intervals are dilated.   That is, an observer measures the time intervals to be increased on a clock

GRAPH 8-3

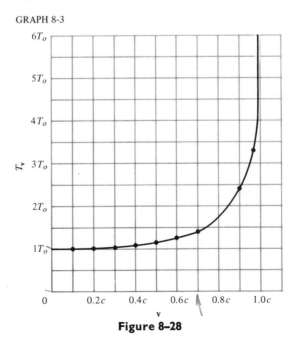

v

Figure 8–28

moving relative to him. If the time interval between ticks on a clock is increased, then the clock runs slow.

It is reasonably easy to understand why moving clocks run slower if we consider a fairly simple time-keeping device. Suppose that the clock in a spacecraft consists of a rod with a flashing light at one end and a mirror at the other end. A unit time interval ("second") in the spacecraft is the time required for a given flash to travel the length of the rod to the mirror, be reflected, and return the length of the rod to the starting point. Let the spacecraft be moving with speed v relative to an observer on Earth. If the rod is perpendicular to the direction of motion, length contraction will not occur.

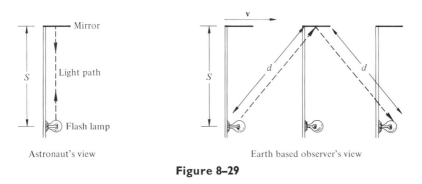

Astronaut's view   Earth based observer's view

Figure 8–29

Now compare the astronaut's view of this clock with the Earth-based observer's view. The astronaut measures the time interval $t = 2\ s/c$ where s is the length of the rod and c is the speed of light. However, since the rod is moving with speed v relative to the man on Earth, he sees the light follow a zigzag path $d = \sqrt{s^2 + (vt)^2}$, which is obviously longer than s. He therefore concludes that the unit time interval is $t' = 2\ d/c$. Since $d > s$, then $t' > t$. Because there is a longer time between "ticks" on the clock as seen from Earth, it runs slower for the earth-based observer than for the astronaut. Notice that the key to time dilation in this analysis is that both observers measure the same, finite speed of light.

Although the analysis is not as straightforward for more complex clocks, any time-measuring device that is moving relative to an observer will seem to that observer to run slowly. The time-measuring instrument might even be a man's pulse rate. If a man's heart and other bodily functions are slowed, then the time he takes to age is also slowed. It is true, then, that a space traveller ages more slowly than the people he left back on Earth. If someday a rocket is developed that will propel a spacecraft at 0.87 the speed of light, you could travel at this speed for 20 years as measured from earth and only age 10 years! This is the basis of the "twin paradox," in which the astronaut returns from a space journey and finds that he is younger than his twin sister, who remained on Earth.

Time dilation, however, is more real than some future high-speed space journey. There are laboratory experiments that verify the predictions of the theory of time dilation. One such experiment involves the lifetime of pi mesons, elementary particles that are unstable and that decay into other particles. The mean life of these mesons, measured by an observer at rest relative to them, is about 2.5×10^{-8} seconds. These particles can be accelerated to a speed of 0.9 the speed of light. From Graph 8–3, what lifetime do you predict for these particles if they are moving at $0.9\ c$? The value determined from the graph is about 5.7×10^{-8} seconds. The experimental result is that these particles travel, on the average, a little over twice as far before decaying as they should if their lifetimes are 2.5×10^{-8} seconds. Therefore, they must have lived a little over twice as long, or something over 5×10^{-8} seconds.

Before we consider the effects that relative motion has on mass, let us attempt to answer a question that is probably in the back of your mind. Is a moving stick *really* shorter? Does a moving clock *really* run slower? The answer depends on what you mean by *really*. Length contraction and time dilation occur through the act of making measurements. To the physicist, what is real is what is measured. Therefore, if a moving rod is measured as shorter than it is at rest, then it is *really* shorter.

Mass Increase.   Consider a spacecraft in free space and therefore in a frictionless situation. Suppose that engineers have developed a rocket for this spacecraft that will fire, and thus exert a constant force on the spacecraft, for many years. What is the final state of the spacecraft? Your analysis, following Newton, is that a constant force acting on the mass of the spacecraft (including the fuel which is decreasing) results in an acceleration $\mathbf{a} = \mathbf{F}/m$. Thus, the spacecraft moves faster and faster, with the only limitation being the length of time the rocket fires. But Einstein

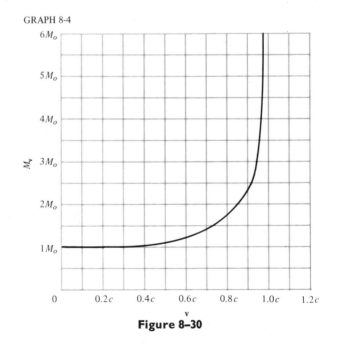

GRAPH 8-4

Figure 8–30

says that this analysis is incorrect because you have neglected the increase in mass of the spacecraft as it gains speed, as shown in Graph 8–4.

The constant force of the rocket increases not only the speed but also the mass of the spacecraft. Since the product of mass and speed (or more correctly, velocity) is momentum, the constant force increases the momentum of the vehicle. As the speed approaches the speed of light, more and more of the force goes into increasing the mass, and thus the spacecraft is accelerated less and less. In this situation, we can appreciate the true meaning of the increase in mass with speed: As the speed of an object approaches the speed of light, its resistance to change in motion (inertia, which is measured by mass) increases. The fact that the mass of an object approaches infinity as its speed approaches the speed of light suggests that an infinite force is required to accelerate a spacecraft (or any other body) to the speed of light. Therefore, the speed of light represents an upper limit for the speed of a material body.

There is another, and in a sense more significant, consequence of the increase in mass with speed. With the knowledge of a little algebra (specifically the binomial theorem), we can write an approximation for $m_v = m_0/\sqrt{1 - v^2/c^2}$ as $m_v \approx m_0 + \frac{1}{2} m_0 v^2/c^2$. If the speed v of the object is small relative to the speed of light, then this is an excellent approximation. Now let us assume that $m_v = m_0 + \frac{1}{2} m_0 v^2/c^2$ and multiply both sides of the equality by c^2. We obtain $m_v c^2 = m_0 c^2 + \frac{1}{2} m_0 v^2$. We recognize $\frac{1}{2} m_0 v^2$ as the classic kinetic energy of an object of mass m_0 moving with speed v. But what of the other two terms? Let us assume, with Einstein, that $m_v c^2$ is the total energy of the object; then $m_0 c^2$ represents what we might call "rest energy." We can rewrite our approximation as $m_v c^2 - m_0 c^2 = \frac{1}{2} m_0 v^2$, or $(m_v - m_0)c^2 = \frac{1}{2} m_0 v^2$ or $(\Delta m)c^2 = \frac{1}{2} m_0 v^2$. But the term on the right is the classic kinetic energy E. Therefore, this relation suggests that, associated with a change in mass Δm, there is an equivalent change of energy $\Delta E = (\Delta m)c^2$. Or, put more simply, any mass m has associated with it an equivalent amount of energy mc^2, where $c = 3.00 \times 10^8$ m/sec. Thus, by means of the equivalence of mass and energy resulting from the special theory of relativity, the law of the conservation of energy and the law of the conservation of mass are united into a single law of the conservation of mass-energy.

The mass-energy equivalence is related to the increase in resistance to change in motion with increasing speed of material bodies. It is one of the most profound and far-reaching results of the special theory of relativity. It is most widely known as the basis for the release of energy in the nuclear reactions of fission and fusion. The fact that this concept is part of the everyday practice of people who work with particle accelerators is not so generally appreciated. For example, in particle accelerators, electrons are accelerated to such speeds that they become as massive as protons (a 2000-fold increase in mass). Electrons and positrons (antielectrons) routinely annihilate each other in research laboratories and produce energy equivalent to the sum of their masses times the square of the speed of light.

The special theory of relativity is special in the sense that the only kind of relative motion to which it applies is the case of constant velocity. Thus, special relativity does not include motion in which one system is accelerated relative to the other. In order to include all kinds of relative motion, Einstein proposed the general theory of relativity. Since the fundamental accelerating mechanism in the Universe seems to be gravitation, general relativity is concerned with gravitation and the mass with which it is associated. The theory is based on a single postulate called the principle of equivalence, which says that there is no difference between the effects produced by matter (gravitation) and those produced in accelerated frames of reference (inertia). The properties of space (geometry, curvature, and so on) are thus determined by the distribution of matter in space. Space is said to have a curvature that is proportional to the local density of matter.

Einstein suggested three astronomical tests of general relativity: (1) Light passing close to a massive body should be deflected. (2) The wavelength of light leaving a massive body should be lengthened. (3) The motion of planets should be altered (advance of perihelion). All three of these predictions have been verified by observations, although in some cases the experimental data are not as conclusive as we would like them to be.

As we view physical reality with regard to relativity, we are forced to give up our preconceived notions regarding absolute space, time, and mass. The precise description of an object, whether it be electron or spacecraft, depends on how fast it is moving relative to the observer describing it. The faster it moves, the more massive it becomes, the shorter it becomes, and the slower its time flows. Events that are separated in space are also separated in time. Our preconceived ideas of the simultaneity of events are called into question. Even our common-sense ideas about space are challenged by the thought that the properties of space depend upon the amount and distribution of matter in the Universe.

Relativity has revolutionized man's understanding of the Universe as probably no other theory has. However, it has been revolutionary in the best sense of that word. It has provided man with a clearer, more accurate, more complete picture of physical reality. It is interesting that the theory of relativity has given us some absolutes in the physical world: the speed of light and the concept that the laws of physics are *independent*, not *relative*.

EXERCISES

1. The general wave relation $v = fL$ becomes $c = fL$ for electromagnetic waves. Use this equation to find the wavelength of an AM radio station that broadcasts on a frequency of 900 kilohertz.

2. What is the frequency of light waves to which the eye is most sensitive if the wavelength for maximum sensitivity is 5.5×10^{-7} meters?

3. What is the speed of light in a glass lens that has an index of refraction of 1.5?

4. White light illuminates a sweater and the sweater appears to be blue. Explain what happened to the colors in the white light.

5. You have probably had the experience of reaching into water to pick up an object, only to find that the object is not where it appears to be. Explain this experience with the refraction of light waves by drawing a diagram of the situation.

6. Draw a ray diagram to scale and thus determine graphically the position and character of the image formed by a diverging lens of focal length F for an object at a distance from the lens of: (a) $1/2$ F, (b) 1 $1/2$ F, and (c) 3 F.

7. Draw a ray diagram to scale and thus determine graphically the position and character (real or virtual; erect or inverted; magnified, diminished or same size) of the image formed by a converging lens of focal length F for an object at a distance from the lens of: (a) $1/2$ F, (b) 1 $1/2$ F, and (c) 3 F.

8. Draw a ray diagram to scale and thus determine graphically the position and character of the image formed by a converging mirror of focal length F for an object at a distance from the mirror of: (a) $1/2$ F, (b) 1 $1/2$ F, and (c) 3 F.

9. Draw a ray diagram to scale and thus determine graphically the position and character of the image formed by a diverging mirror of focal length F for an object at a distance from the mirror of: (a) $1/2$ F, (b) 1 $1/2$ F, and (c) 3 F.

10. Explain the appearance of a distant street light when seen through screen wire or a piece of finely woven material, such as an umbrella.

11. Why do you not observe interference effects from two lamps in the same room?

12. Explain why colors are seen when white light is reflected from a soap film.

13. Suppose you suspect that the sunglasses you are about to purchase are not really Polaroid. How could you determine for sure whether they are Polaroid or not by stepping outside the store (assuming the sun is shining)?

14. The distance from Earth to the moon is determined most accurately by measuring the time taken for a light pulse to travel from Earth to moon and to be reflected back to Earth. In such an experiment, the round-trip time interval is found to be 2.6 seconds. What is the Earth-moon distance?

15. The circumference of the Earth at the equator is approximately 40,000 kilometers. How many times would a light pulse circle the Earth at the equator in 1 second?

16. Suppose that you are in a vehicle that has no windows and that is soundproof, so that you have no contact with the outside. Is there any experiment you can perform to determine (a) whether you are at rest or are moving with constant velocity? (b) Whether or not you are rotating? (c) Whether or not you are undergoing linear acceleration?

17. Refer to Graph 8–4. How fast would you have to travel in order for the mass of your body to double, as measured by an observer at rest?

18. Refer to Graph 8–3. You and your clock are traveling at three fourths the speed

of light. If 1 hour elapses on your clock, how much time would an observer at rest observe to elapse on your clock?

19. Discuss the problem of accelerating an electron to the speed of light.

20. According to the theory of general relativity, does light from a distant star really travel in a straight line to Earth? Explain.

21. In Chapter 3, some of the problems challenging man's dreams of exploration beyond the solar system were mentioned. Re-evaluate these problems in the light of the special theory of relativity. Are length contraction, time dilation, and mass increase advantages or disadvantages (or neither) in a space trip requiring tens of years? Is the mass-energy equivalence $E = mc^2$ related to this problem?

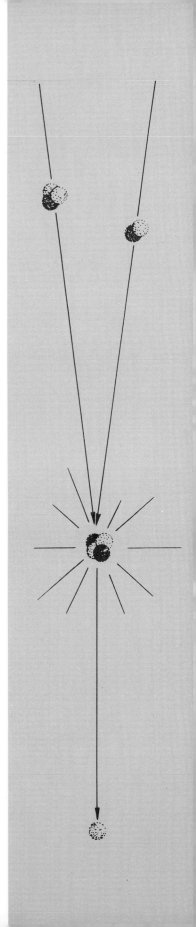

INSIDE
THE ATOM

THE EXTRANUCLEAR ATOM

In this chapter, we will be investigating the interior of the basic building block of all matter—the atom. In our attempts to understand the composition, structure, and reactions inside the atom, we are handicapped by the fact that we cannot physically look inside to study it. How can we analyze the interior of a black box into which we cannot even see? One method is to study what comes out of the box. Another is to study what happens to things that go into the box. We will thus attempt to construct a reasonable interior of the atom by observing things entering and leaving it. This is an adventure in model building—probably the most difficult exercise in model building that man has attempted. The challenge is so great that after 75 years the model is still incomplete. Much of the fundamental research in physics today is concerned with developing a better model of the atom and its nucleus.

THE QUANTUM CONCEPT

Before we begin constructing a model of the atom, we need to consider a fundamental concept upon which the model is based. This concept, which was introduced into physics in the early part of this century, has had a revolutionary impact on physics and therefore on man's understanding of physical reality. Perhaps a bit of history will put this revolution in perspective.

You will recall that in the early years of the nineteenth century, the wave nature of electromagnetic energy (including light) was firmly established through experiments involving diffraction and interference. By the 1880's, the theory of electromagnetic radiation was essentially complete. No sooner had Maxwell finished this monumental task than some experiments with electromagnetic radiation gave results that were difficult to explain in terms of waves. Outstanding among these was the theoretical analysis of the radiation of heat from bodies at high temperatures.

Many attempts were made to explain the energy distribution as a function of frequency for electromagnetic radiation emitted from hot objects. No progress was made until Max Planck made his revolutionary proposal in 1900. Planck suggested that the failure to explain these experiments was rooted in the basic assumption that radiation is a continuous (wave) phenomenon. Planck proposed that the energy is radiated in discontinuous bursts called quanta, and that the amount of energy contained in each quantum is proportional to the frequency of the radiation. Using this new idea, Planck was able to explain the energy-frequency relationships that were observed in the experiments.

Just what is a quantum of energy? In everyday language, a quantum is nothing more than a bundle or chunk of energy. Although the idea of chunks of radiant energy seems very strange, we are quite familiar with other manifestations of quantized energy. After all, particles and electric charges (both forms of energy) are quantized. Therefore, just as we think of a smallest particle (the electron) and of a smallest unit of charge (the charge on the electron), neither of which can be subdivided, so we must also think of a smallest unit of energy (the quantum) that cannot be subdivided. We do not encounter half electrons; neither do we encounter half quanta. The value of the smallest bundle of radiant energy is a constant times the frequency of the radiation:

$$E_R = hf$$

where h is a constant, now called Planck's constant, which is found experimentally to have the value 6.63×10^{-34} joule-seconds.

An idea as radical as Planck's was not easily accepted. Planck himself preferred to think of the quantization as being associated with the emission process and not with the radiation itself. This concept had to be clarified and confirmed through its application to another experiment. This extension of the quantum concept was provided by Albert Einstein in 1905 (the same year in which he published his special theory of relativity) through his use of the quantum theory to provide an explanation for the photoelectric effect.

In the photoelectric effect, electrons are ejected from the surface of a material by the energy of electromagnetic radiation falling on the surface. A simplified experimental arrangement is shown in Figure 9–1. The ejected photoelectrons are detected by attracting them to a plate held at a positive potential by a battery, and measuring the resulting current in the circuit with an ammeter. In performing the experiment, there are two variables we can control—the intensity of the light and the frequency of the light. The two dependent variables we can observe are the number of electrons ejected (by measuring the current) and the maximum kinetic energy of the ejected electrons (by making the polarity of the plate negative

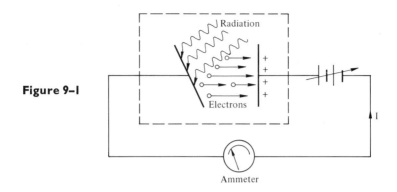

Figure 9-1

and varying its potential). The number of electrons ejected essentially agrees with classic expectations, but attempts to interpret the kinetic energy data according to classic expectations encounter serious difficulties. The experimental results of variations of maximum kinetic energy with intensity and frequency are shown in Figures 9–2 and 9–3.

Analyze each graph carefully, asking in particular for each whether or not the results are consistent with classic (pre-quantum) physics. What about Graph 9–1? As the light gets brighter, the maximum kinetic

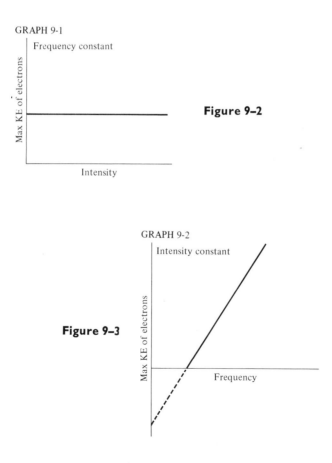

GRAPH 9-1

Frequency constant

Max KE of electrons

Intensity

Figure 9–2

GRAPH 9-2

Intensity constant

Max KE of electrons

Frequency

Figure 9–3

energy of individual electrons remains the same. This is a little difficult to explain. Analysis suggests that a brighter light would not only eject more electrons, but would also impart more energy to the ejected electrons. Now consider Graph 9–2. Our analysis of this graph is that the maximum kinetic energy of ejected electrons is a function of the frequency of the light. This presents a real problem, because classically the energy of a wave (and light is a wave phenomenon) depends on the amplitude of the wave, not on its frequency. Furthermore, you observe that the curve does not pass through the origin, but intercepts the frequency axis at a frequency well above zero. This implies the existence of a minimum frequency below which the electrons have no kinetic energy and therefore are not ejected. There is one other result of the photoelectric experiment that has no classic explanation, and that is the fact that when light strikes the surface, photoelectrons are emitted immediately, no matter how low the intensity of the light is.

If radiation is a continuous wave phenomenon, why does the energy of the electrons not depend on the intensity (energy) of the radiation? Why does the energy increase with the frequency? Why are electrons ejected immediately, regardless of how low the intensity is? Why is there a frequency below which no electrons are ejected, no matter how intense the light is?

Einstein proposed (following Planck's lead) that the radiation is not continuous, but rather is emitted in bursts (quanta), and that the energy of each burst is a function of the frequency. Therefore, an electron either receives a quantum of energy hf or it receives no energy at all. It also follows from this concept that there exists an energy, and therefore a frequency, below which no electrons would be ejected. However, if a quantum contains enough energy to eject an electron, then that electron should be emitted immediately, no matter how few quanta (how low the intensity is) are striking the surface. The kinetic energy of an ejected electron, then, is the energy of a quantum of light hf less the energy E_0 the electron loses in escaping from the surface of the material. This is expressed in symbol form as $KE = hf - E_0$. Relating this equation to

GRAPH 9-3

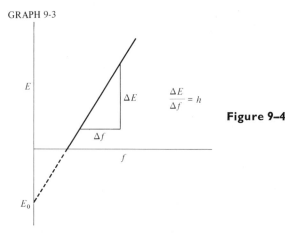

$$\frac{\Delta E}{\Delta f} = h$$

Figure 9–4

Graph 9–3 shows that the slope of the curve is Planck's constant h and that the negative intercept on the vertical axis is the work E_0 required to free an electron from the material. Therefore, Einstein demonstrated in his theoretical analysis of the photoelectric effect that radiation itself is quantized and that the energy of a quantum is hf. A quantum of light is called a *photon* (light particle).

Although the photon model of light has answered many of our questions regarding physical reality, it has also raised a very basic question. A photon has some characteristics of a particle—it is discrete and finite. But light is also a wave, as demonstrated by diffraction and interference experiments. Therefore, it appears that light is both a particle and a wave. But that is a contradiction, since a particle is finite and discrete and a wave is infinite and continuous. Is light a particle or is it a wave? The best answer we can give to this is that in some experiments light behaves like a particle and in other experiments it acts like a wave. One interesting fact is that in no experiment does light exhibit both characteristics simultaneously. Physicists consider wave theory and particle theory as complementary rather than contradictory. A complete understanding of physical reality is possible only through the acceptance of both theories. It may be that there is some very basic undiscovered concept of physical reality that would clarify this apparent contradiction.

Since waves sometimes manifest themselves as particles, do particles ever exhibit wave characteristics? Arthur Compton showed in the early 1920's that the scattering of electromagnetic waves by free electrons implies that a photon has a momentum of magnitude h/L, where L is the wavelength of the waves. This concept led Louis de Broglie to suggest in 1924 that the motion of a particle has associated with it a wavelength that is equal to Planck's constant divided by the particle's momentum. The momentum of a particle is just the product of its mass and velocity. Therefore, any particle has a wavelength $L = h/mv$.

This hypothesis was verified experimentally in the late 1920's. A beam of particles (electrons) was passed through a thin layer of crystals that functioned as a diffraction grating for the electrons. The electrons, after passing through the crystals, produced a diffraction pattern similar to the pattern observed in the diffraction of light. The fact that the diffraction interaction is strictly a wave phenomenon implies that in this experiment the electrons exhibited wave characteristics. Since h is very small (on the order 10^{-34} joule-second), an object of ordinary mass and velocity has an extremely short wavelength.

Einstein's explanation of the photoelectric effect left little doubt of the reality of quanta. In the years that followed, the photon model of light was used successfully to explain other phenomena, most important of which were the spectra emitted from atoms and the interaction between electromagnetic radiation and free electrons (Compton effect). However, it was not until the late 1920's that a complete quantum theory was worked out. The embodiment of this theory was the wave mechanics, or quantum mechanics, initiated by de Broglie and developed primarily by Erwin Schrödinger and Werner Heisenberg. Wave mechanics provides a mathematical model of the physical world. The discreteness of natural phenomena is not assumed in the model; instead, quantization appears in the solutions of the mathematical equations.

One of the most significant features of this model is the conclusion that, in describing physical reality, strict determinism must be replaced by probability. Wave mechanics implies that ultimately all we can do is characterize a system in a statistical sense and then predict the future states of the system in this same statistical sense. Wave mechanics, then, is more properly a model that predicts the probability of events rather than a model that provides a pictorial representation of the situation. Today, wave mechanics is applied routinely to the analysis of all subatomic phenomena; in fact, it is the basis for our understanding of atomic and nuclear structure and reactions.

THE NUCLEAR MODEL OF THE ATOM

What is the inside of an atom like? Suppose we begin this exercise in model building by assuming that there are charged particles inside the atom. This seems to be a reasonable assumption, since the existence of electrons is well established and there are experiments that lead us to believe that these negatively charged particles are found within the atom. So, let us assume that at least one atomic constituent is the electron. But, since atoms as a whole normally are electrically neutral, the model must also include positively charged particles, in order that the net charge of the atom be zero. So now we have positively charged particles and negatively charged particles in our atom. The question now is, "How are these particles distributed?" Is the distribution like an equal number of black and white marbles thoroughly mixed in a box? Or are the negatively charged particles embedded in a positively charged material like raisins in a pudding? Or is one charge concentrated at one point and the other charge concentrated at another? You can think of still other charge distributions. Thus there are several possible models. Our problem is to choose the best model from several possible ones. What we need is an experiment that will give us some clue regarding the arrangement of the charges.

Suppose that a ball of fluffy, white cotton the size of a basketball is placed on a table near where you are sitting. You are told that the ball of cotton contains hard metallic objects and are asked to determine the size of these objects and how they are distributed in the cotton without leaving your seat. One way you might approach this problem is by shooting bullets into the ball and studying the pattern made by the bullets in a sheet of paper suspended behind the cotton. If there are numerous small metal balls in the cotton, you expect the bullets to be deflected only slightly from their paths, forming a fairly random pattern of holes in the paper behind the cotton. However, if the hard material is a single metal sphere at the center of the cotton ball, then a bullet will not be deflected at all unless it happens to hit the metal sphere. If it does hit the sphere, it is likely to suffer a large deflection. If many bullets are fired, some regularity in the pattern would be expected, with a shadow containing no bullet holes behind the metal sphere. You are right in thinking that is a hard way to find out what is inside a ball of cotton. However, it does suggest a technique that might allow us to probe the inside of an atom.

Let us now replace the ball of cotton with an atom and consider what other modifications we will need to make in our investigation. It is obvious that the bullets must be much smaller than the object being probed—an atom, in this case. Since we are interested in studying the charge distribution in an atom, the bullet, in addition to being small, must also be charged. Fortunately, there are bullets that are charged and smaller than atoms, such as electrons and protons. As it turns out (for reasons we will not discuss here), the most practical bullets are alpha particles. Alpha particles are helium nuclei and thus have a charge of $+2$ electronic charges. Alpha particles are readily available, and the only problem in the experiment is getting a collection of atoms thin enough for the alpha bullets to penetrate. This is most easily done by using a thin metallic foil, such as gold. A cross-sectional view of the experimental arrangement is shown in Figure 9–5.

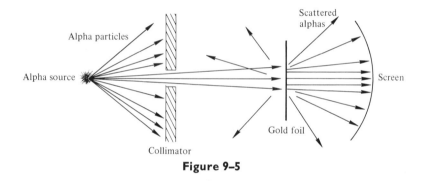

Figure 9–5

The results of the experiment are that most of the particles pass through the atoms with no deflection. Some particles, however, are deflected at rather large angles, and occasionally a particle undergoes a very large deflection, even as much as 180°, returning in the direction from which it came. Our interpretation of these results is based on the principle of interactions between charged particles. We expect the deflection of alpha particles by electrons to be negligible, since the alpha particle is some 7000 times as massive as an electron. Thus, we assume that the alpha particle deflections are caused by the repulsive force between the alpha particles and positive charges in the atom.

A careful study of the pattern of alpha particle "holes" in the screen suggests that the positive charges do not permeate the space of the atom, but that they are highly concentrated along with most of the mass of the atom near the center of the atom. These then are the chief features of the nuclear model of the atom: (1) All the positive charges of an atom are concentrated in an extremely small space called the nucleus, (2) almost all the mass of an atom is also in the nucleus, (3) the negative charges of an atom are in the form of electrons, which are relatively very far from the nucleus, and (4) an atom is mostly empty space.

The data from this alpha-scattering experiment are so well defined that they lead to a value between 10^{-15} meter and 10^{-14} meter for the diameter of the nucleus of an atom. Other experiments give the atomic diameter at approximately 10^{-10} meter. Thus, the diameter of an atom is approximately 100,000 times the diameter of its nucleus! If the nucleus were scaled up to golf-ball size, the atom of which it is the nucleus would be nearly two miles in diameter and almost all the mass would be in the golf ball.

SPECTRA

The model of the atom, as we have developed it thus far, pictures most of the mass and all the positive charges as being concentrated in a very small volume at the center of the atom. This nucleus is surrounded by electrons equal in number to the positive charges of the nucleus. Are you satisfied with this model? Is our picture of an atom complete?

How do you visualize the electrons that are around the nucleus? Since particles sent into an atom give us no information regarding the arrangement of electrons, we must look for some other indication of the electron distribution. Is there any energy coming out of the atom that might contain some clues as to the structure of the atom outside the nucleus? An energy source that bears investigating is a glowing body. If we can establish that this energy comes from inside the atom, we might be able to recover some information regarding atomic structure from it.

You have probably seen charcoal briquets heated in a fire. The charcoal, of course, emits heat energy, but the fact that it glows suggests that it also emits energy in the form of light. Furthermore, the light emitted changes color (and therefore frequency, and therefore energy) from red to white to maybe even blue-white. Energy implies information, but how do we extract information from this light energy? We need to analyze the energy carefully, possibly by breaking it down into its component parts.

Fortunately, in the case of light we have a technique already perfected for doing just that. We have demonstrated that either a prism or a diffraction grating can be used to display the components of light. Such a spectrum of light from an incandescent solid proves to be rather disappointing, since the display is a continuous array in which one color gradually fades into the next color. A similar investigation of the light from an incandescent, low-pressure gas, on the other hand, turns out to be very revealing.

A continuous spectrum is produced by a gas under high pressure. However, the spectrum of light from an incandescent gas under low pressure is not continuous at all, but rather consists of an orderly arrangement of bright lines of different colors separated by dark spaces. In order for a rarefied gas to emit radiation, energy must be supplied to it. A much easier way of exciting a gas to incandescence than by heating it is to increase its energy by producing an electrical discharge in the gas. Such a discharge tube consists of a sealed glass tube containing gas at a given pressure. The tube has electrodes at either end, to which a potential difference of a few thousand volts is connected. The resulting electrical discharge excites the gas to incandescence. The spectrum of this light can be investigated by using a spectrometer. A spectrometer consists of a lens to make the light rays parallel, a prism or diffraction grating to display the components, a telescope to study the spectrum in detail, and a scale to permit accurate measurements to be made.

Upon observing the spectra of various gases through either a spectrometer or a diffraction grating held near the eye, you will see patterns of bright lines on a dark background. If a calibrated spectrometer is used, the wavelength of the light producing each line can be determined. A few representative bright-line spectra are shown in Figure 9–6.*

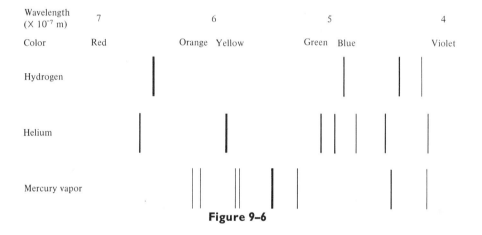

Figure 9–6

Careful observation of spectra resulting from rarefied gases leads us to the following generalizations: (1) Excited gases emit radiation, (2) their spectra consist of discrete, bright lines, (3) a given element always produces the same pattern of lines, (4) different elements exhibit different patterns, and (5) there seems to be some regularity in the spacing of lines, particularly for hydrogen. Since whatever information this energy may contain is pertinent to our present problem of atomic model building only if the energy comes from inside atoms, generalizations (3) and (4) are very important to us. Atoms are the building blocks of elements. A given element always produces the same pattern, and no two different elements produce the same pattern. Therefore, the conclusion that this energy originates inside atoms is almost inescapable. Since the spectrum of an element is unique, a spectrum is the "fingerprint" of a given element. Thus, elements can be identified from a study of their spectra. This

---

* It is recommended that Experiment 13 in the lab manual be done at this time.

technique is particularly important in astronomy, where elements in the sun and other stars are identified from spectral analysis.

How, then, do we proceed to analyze this energy coming from inside atoms? The facts that the hydrogen spectrum is the simplest of all spectra and that it shows the most noticeable regularity of line spacing lead us to begin our investigation with the spectrum of this element. Many years before spectra were associated with the internal structure of atoms, work was being done on the characteristics of the spectra themselves. In the 1880's, Balmer made a careful study of the wavelengths of the lines in the hydrogen spectrum, and discovered a mathematical relation that gives the wavelengths of the four lines in the visible spectrum. The empirical formula for these lines (now called Balmer series) is:

$$1/L_n = R(1/2^2 - 1/n^2).$$

In this formula, L_n is the wavelength of the light producing a particular line, R is a constant that has the value 1.10×10^7 per meter, and n is an integer having the value 3 for the red line, 4 for the blue-green line, 5 for the blue line, and 6 for the violet line. If integers greater than 6 are substituted in the formula, the resulting wavelengths are too short to be seen. These lines in the ultraviolet region of the electromagnetic spectrum were subsequently detected. This formula thus gives values of a series of lines that get closer together with decreasing wavelengths. The integer 2 is used in the denominator of the first term in the parentheses because substitution of integers such as 1 and 3 in the formula gives wavelengths in either the infrared or the ultraviolet ranges. Thus, there are other series of lines that are not visible. The existence of these lines was also verified experimentally.

Before we attempt to relate the hydrogen spectrum to the structure of the hydrogen atom, there is one other significant feature of the spectrum that we should notice. The lines in the hydrogen spectrum, and in the spectra of all elements, are discrete. A particular line begins and ends sharply, there is a gap with no lines, and then another isolated line begins. This same discreteness is also evident in the formula for the wavelengths, in that only integral values of n result in observed lines. There is a line for $n = 3$ and another for $n = 4$, but no line corresponding to $n = 3.5$. This reminds us of the concept of quantization that we developed earlier. We will need to keep this fact in mind as we build our model of the hydrogen atom.

A MODEL OF THE HYDROGEN ATOM

Now that we are reasonably sure that the light that produces the lines in the hydrogen spectrum has a subatomic origin, how do we relate this information to the structure of the atom? First, we need to know from what part of the atom the energy comes. Does light originate in the nucleus, or in the extranuclear atom?

Chemistry provides us with a significant clue toward establishing the source of light in the atom. One of the basic assumptions in chemistry is that when atoms interact to form molecules, the reaction involves the extranuclear atom only. In fact, the chemical properties of an element are determined by the electron distribution in the atoms of that element. Now, as can be demonstrated, changing the electron distribution either by removing an electron from an atom (ionizing) or by combining the atom with another to form a molecule changes the bright-line spectrum produced by that element. The logical conclusion drawn from these experiments is that the light energy that produces the hydrogen spectrum, or any other, originates in the extranuclear electrons. What model of electron structure do we arrive at from an analysis of the hydrogen spectrum?

In 1913, on the basis of the evidence available, Neils Bohr proposed the idea that the hydrogen atom consists of a single electron in a circular orbit about a positively charged nucleus. The orbit is the result of the electron's tendency to continue in a straight line (Newton's law of inertia) and the electrostatic attraction (Coulomb's law) of unlike charges. Thus, basically, the Bohr model of atomic structure pictures the hydrogen atom as a miniature sun-planet system. However, the force that supplies the centripetal force to keep the electron in orbit is electrostatic rather than gravitational. The magnitude of this coulomb force between the nucleus and the electron is $F = kQ_1Q_2/r^2 = k\,e \cdot e/r^2$, where k is a constant, e is the magnitude of the charge on the electron and on the hydrogen nucleus, and r is the distance between the nucleus and the electron (see Chapter 4). The centripetal force, you will recall from Chapter 3, is given by $F_c = mv^2/r$, where m is the mass of the electron, v is its linear (tangential) speed, and r is the distance between nucleus and electron. If the electron is in a stable, circular orbit then $F_c = F_Q$, or $mv^2/r = k\,e^2/r^2$, which simplifies to $mv^2 = k\,e^2/r$.

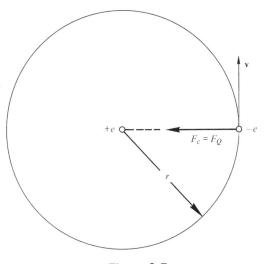

Figure 9–7

The electron and the nucleus thus constitute a bound system. Since an energy analysis proved fruitful when we analyzed a planet-sun system, let us consider the energy of the electron-nucleus system. The total mechanical energy of the system is the kinetic energy of the electron (assuming that the nucleus does not move) plus the potential energy of the electron-nucleus system: $E = KE + PE$. But $KE = \frac{1}{2} mv^2$, where m is the mass of the electron and v is its speed. The PE is (from Chapter 4) the area under the force vs. displacement curve and in symbol form is given by $PE = -ke^2/r$. Therefore, the total energy of the system is $E = \frac{1}{2} mv^2 + (-ke^2/r)$. But, from the analysis in the preceding paragraph, $mv^2 = ke^2/r$; therefore, $\frac{1}{2}(mv^2) = \frac{1}{2} ke^2/r = KE$. Thus $E = KE + PE = \frac{1}{2}(ke^2/r) - ke^2/r = -\frac{1}{2} ke^2/r$. The variation of KE, PE, and E with electron-nucleus distance is shown in Graph 9-4. The negative total energy of the electron-hydrogen nucleus system is analogous to the negative energy of the bound sun-planet system discussed in Chapter 3. It means that the atom is a bound system and that energy equivalent to $\frac{1}{2} ke^2/r$ must be supplied to the system to remove the electron to such a distance that it is no longer bound to the nucleus. As the radius r of the orbits increases, E increases and approaches zero as r becomes very large, as shown by the total energy curve in Graph 9-4.

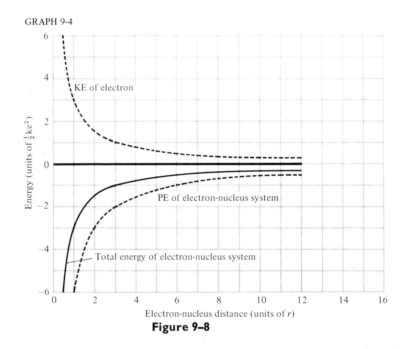

GRAPH 9-4

Energy (units of $\frac{1}{2}ke^2$)

KE of electron

PE of electron-nucleus system

Total energy of electron-nucleus system

Electron-nucleus distance (units of r)

Figure 9–8

So far, the Bohr model of the hydrogen atom is identical to the circular solar system model, with the gravitational force replaced by the electrostatic force. But there is a fundamental difference between the two systems. The electron is a charged particle moving in an orbit, and therefore, according to the ideas developed in Chapter 8, it radiates electromagnetic energy. How can the electron remain in a stable orbit if it is losing energy and therefore causing the energy of the system (which depends on the radius of the electron orbit) to change?

To overcome this difficulty, Bohr proposed the quantum concept: that only orbits of certain radii are possible; that is, the electron orbits are quantized. The quantized orbits correspond to quantized energy levels for the electron. Thus, the electron only radiates when it changes energy levels from an orbit of a larger radius to an orbit of a smaller radius. There is a lowest energy level (ground state) corresponding to the smallest allowed radius. If the electron is in any orbit of larger radius than this, it has above normal energy, and the atom is said to be in an excited state.

You may be wondering how the hydrogen spectrum fits into this model. For one thing, it has already contributed the idea of quantized orbits. But let us pursue Bohr's idea of energy levels a little further. Consider a hydrogen atom with its electron in its lowest energy level (orbit of smallest radius). In order for it to go to a higher energy level, some energy must be supplied from the outside, possibly in the form of heat or an electrical discharge. An electron thus raised to a higher energy state by external energy finds itself in an abnormal condition with an excess of energy. But in order to return to its normal, lowest energy condition, the electron must get rid of some energy, and it releases this energy as a quantum of electromagnetic radiation. Therefore, E (higher) $- E$ (lower) $= hf$. Thus, a spectrum line that we observe from excited hydrogen is emitted when the electron changes from a higher to a lower energy state. The wavelength of this line can be determined with a spectrometer,

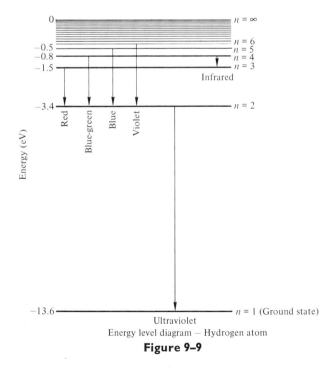

Energy level diagram — Hydrogen atom

Figure 9–9

and the energy of the photon can be calculated from the relation $E = hf = hc/L$. But this energy is just the difference between two energy levels in the atom. By careful study of all of the lines in the spectrum, we can construct a model of the possible energy levels for the electron in a hydrogen atom, as shown approximately to scale in Figure 9–9.

Energies at the atomic level are so extremely small (for example, the energy involved in electron orbit transitions is of the order of 10^{-18} joule) that it is convenient to define an energy unit for use with atomic phenomena. This unit is the *electron volt* (eV), which is defined as the kinetic energy an electron acquires in being accelerated by a potential difference of one volt. One electron volt is equivalent to 1.60×10^{-19} joules.

The values of n that number the energy levels are referred to as *quantum numbers*, or more correctly, *principal quantum numbers*, since additional numbers are required for a more detailed model. The relationship between allowed energies and quantum numbers is $E_n = -(1/n^2)E_1$, where E_1 is the lowest energy state possible. Thus, the allowed energies for the electron-nucleus system for the hydrogen atom are $-E_1$, $-E_1/4$, $-E_1/9, \ldots$. The energies are negative because this is a bound system. Energy must be added to an electron in a lower level to raise it to a higher level. The energy required to remove an electron from the lowest level to a distance at which it is no longer influenced by the nucleus (energy of system is zero) is 13.6 eV for hydrogen. This energy is called the ionization energy. An atom from which an electron has been removed is said to be ionized.

As shown in Figure 9–9, the four visible lines in the hydrogen spectrum result from electron transitions to the second energy level. This would represent the 2 in the Balmer formula. The transitions originate at levels 3 (red), 4 (blue-green), 5 (blue), and 6 (violet). These, of course, are the values of n in the Balmer formula. Thus, you see that this model is consistent with spectral analysis.

Consider now a transition from $n = 2$ to $n = 1$. This line represents more energy (higher frequency, shorter wavelength) than violet, and therefore it should be in the ultraviolet region. A transition from, say, $n = 4$ to $n = 3$ represents less energy (lower frequency, longer wavelength) emitted than red, and therefore it appears in the infrared region. Thus, the complete model is based on analysis of visible, ultraviolet, and infrared spectra.

It is important to keep in mind that a given hydrogen atom has only one electron. This electron can be in only one level at a given time, makes only one of the transitions, and thus emits only one particular line in the spectrum. The spectrum consisting of many lines results from innumerable hydrogen atoms, each with an electron making one of the various transitions at a time. A brighter line indicates that more atoms have an electron making the corresponding transition. The fact that some lines are always more intense than others suggests that some transitions are more probable than others. An electron in a higher energy level can generally return to its lowest level by several routes. For example, an electron in $n = 4$ could cascade down to 3, then to 2, and finally to 1, producing three spectral lines on the way, or it could drop directly from 4 to 1, producing one line of higher energy.

Through a mathematical analysis of the forces and energies involved, we can show that the relationship between the radii r of the electron orbits and the quantum numbers n is given by $r \propto n^2$. Thus, we picture the extranuclear hydrogen atom as having an electron in an orbit of smallest radius r_1, where r_1 is approximately 5×10^{-11} meters. The other possible orbits for the electron are given by $r_n = n^2 r_1$. The first four orbits in a hydrogen atom are shown to scale in Figure 9–10.

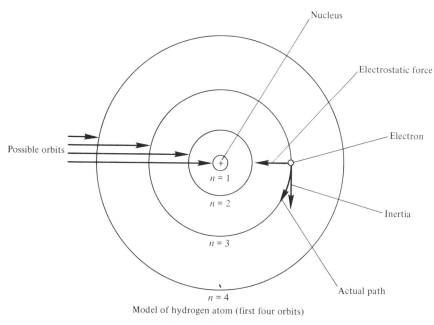

Model of hydrogen atom (first four orbits)

Figure 9–10

Why are electron orbits in an atom quantized? Why are orbits of certain radii not possible? The answer is found in the concept that we introduced earlier, that every mass has associated with it a wavelength $L = h/mv$. An electron, therefore, has a wavelength that depends upon how fast it is moving. An electron will maintain a stable orbit only if the circumference of an orbit is an integral number of the electron's wavelength; otherwise, destructive interference will cause a disruption of the orbit. The circumference of a circular orbit of radius r is $2\pi r$. Therefore, the condition for a stable orbit for an electron of wavelength L is $nL = 2\pi r_n$, where $n = 1, 2, 3, \ldots$. However, $r_n = nL/2\pi$ is not inconsistent with the statement that r_n is proportional to n^2, because L is a function of v, which also depends on n.

Before we attempt to extend this model of the structure of the hydrogen atom to more complex atoms, think for a moment about the new insights we have gained into the nature of light in this analysis. In our consideration of light, we concluded that it has its origin in moving electric charges. However, the question of the source and nature of motion of these charges was left unanswered. We now understand that light has its origin in the orbital electrons of excited atoms and that different colors (photons of different energies) are produced as these electrons make different energy transitions within atoms.

THE ELECTRONIC STRUCTURE OF ATOMS

The remarkable success of the Bohr model of the hydrogen atom in accounting for the observed spectrum impels us to extend the model to other atoms. Unfortunately, our attempts to apply the hydrogen model to more complex atoms meet only with failure. The model does not even account for the spectrum of the next simplest atom, helium. These failures do not mean that the model is incorrect for hydrogen, but rather that it is limited in its scope. The hydrogen atom is a simple, two-body system, whereas all other atoms (at least in an un-ionized state) are multiple-body systems. The analysis of such systems is very complex. For example, in a helium atom, an electron is influenced not only by the positive charge of the nucleus, but also by the other electron, which in turn is affected by the first electron, and so on. Formulation of a model for these more complex atoms requires the invention of additional quantum numbers and the development of a more sophisticated theory. These developments in wave mechanics theory were discussed earlier. This theory presents a mathematical model of the atom but adds little to our visualization of the inside of an atom. When the need for a concrete picture of the interior of an atom arises, we still tend to fall back on the Bohr model of the hydrogen atom and assume that more complex atoms must be something like it. Our faith in the extension of the fundamental ideas of this model to more complex atoms is strengthened by the fact that the spectrum of each element consists of discrete lines. We therefore assume that quantized energy levels exist for the electrons in every atom.

Since the quantum mechanical model is beyond the scope of our development, we seek some acceptable, if perhaps incomplete, understanding of how the atoms of the elements are built up. Suppose we begin with the simplest atom, hydrogen, and construct the elements by adding one electron at a time. The chemical properties of elements as exemplified in the periodic table suggest that each element has one more electron than the preceding element. As we add electrons, we must also add positive charges to the nucleus to hold the electrons in their orbits. In addition, mass must be added—we will assume it is added to the nucleus—in order to have the mass of the atoms increase as we go along.

It turns out that there are two conditions imposed on each added electron. The first condition, which seems quite reasonable, is that an added electron must go to the lowest energy level available in the atom. The second condition is that the added electron cannot have the same quantum state as an electron that is already in the atom. This requirement placed on electrons in an atom is known as the Pauli exclusion principle and simply states that no two electrons in the same atom can have identical quantum properties. Unfortunately, this second condition does not have the common-sense appeal of the first, but nevertheless it is consistent with all theoretical as well as experimental analyses.

Applying the preceding conditions, you would put the electron that is added to hydrogen to form helium into the second energy level. However, in the model it is in the first energy level. Now the question arises as to how two electrons can be in the same energy level and not have identical quantum properties. The answer is found in the fact that electrons have an additional quantum property called *spin* associated with them. Although the concept of spin of elementary particles cannot be developed at this level of treatment, we can understand the results through a somewhat oversimplified model.

Imagine that as an electron orbits the nucleus it also rotates on its axis. Only two directions of rotation are possible—clockwise and counterclockwise. Thus, electron spin is quantized into two discrete states, each of which represents a different energy state. Therefore, the two electrons in the first energy level of helium do not violate the Pauli principle because they differ in spin. However, a third electron cannot be put into this lowest energy state because it will be in a state identical to one of the two electrons already there. The third electron in a lithium atom must go into the second energy level, since it is the lowest energy state available. The next element, beryllium, is formed by adding the next electron to the second level, as expected.

Our scheme of adding electrons is now complicated by the fact that except for the first, all energy levels, or shells, as they are more commonly called, consist of subshells. Electrons in different subshells have different sets of quantum numbers and therefore are not identical. Actually, two quantum numbers (one for the orbital motion and the other for its orientation in space) are associated with the electron subshell configuration. These two values, together with the principal and spin numbers, give each electron a set of four quantum numbers. According to the Pauli principle, no two electrons in the same atom can have the same set of four quantum numbers. Further development of this concept leads to the result that the maximum number of electrons allowed in the second shell ($n = 2$) is eight. The maximum allowed in the $n = 3$ shell is 18, and in general the maximum number of electrons allowed in any energy level n is $2(n^2)$. The existence of subshells implies that in order to develop a more complete model, we must consider the filling of subshells as well as the filling of shells. Adherence to the conditions imposed by nature on the distribution of electrons leads to the model of electronic configurations shown in Table 9–1.

Table 9–1 is based on the idea that a new row is begun each time a shell or subshell is filled with electrons. In the periodic table devised by chemists, elements are listed in order of increasing charge and elements with similar properties are placed under each other (vertical columns). The rows in the table then are periods in which the various properties are

TABLE 9–1   ELECTRONIC CONFIGURATION IN ATOMS.

ELECTRONS IN				ELEMENT				
	H							He
1st shell	1							2
	Li	Be	B	C	N	O	F	Ne
1st shell	2	2	2	2	2	2	2	2
2nd shell	1	2	3	4	5	6	7	8
	Na	Mg	Al	Si	P	S	Cl	A
1st shell	2	2	2	2	2	2	2	2
2nd shell	8	8	8	8	8	8	8	8
3rd shell	1	2	3	4	5	6	7	8
	K	Ca	....				Br	Kr
1st shell	2	2	....				2	2
2nd shell	8	8	....				8	8
3rd shell	8	8	....				18	18
4th shell	1	2	....				7	8
	Rb	Sr	....				I	Xe
1st shell	2	2	....				2	2
2nd shell	8	8	....				8	8
3rd shell	18	18	....				18	18
4th shell	8	8	....				18	18
5th shell	1	2	....				7	8

repeated. The fact that the atomic structure model we have constructed agrees perfectly with the periodic classification of the properties of the elements contributes greatly to our confidence in this particular model of the atom. Elements with similar chemical properties have atoms with remarkably similar electron configurations; therefore, it follows that the properties of an element are determined essentially by the number of electrons in its outer subshell. For example, fluorine, chlorine, bromine, and iodine (each with seven electrons in the outer shell) have similar chemical properties.

Notice that adding an electron to form the next element goes according to plan through argon. The first shell is completed with helium, the second with neon, and the first subshell of the third shell fills at argon. However, irregularities begin to occur at potassium. The electron added to argon to build the model for potassium goes to the fourth shell even though the third shell is not complete (the third·shell can contain $2(3)^2 = 18$ electrons). The explanation for this is found in the overlapping of subshells. The first subshell in the fourth shell has a lower energy value than some subshells in the third shell. The added electron thus goes to the lowest unoccupied energy state.

We might now inquire about the results of removing an electron from an inner shell in a more complex atom. We have seen that when an electron vacancy in an outer shell is filled, infrared, visible, or ultraviolet radiation is emitted. An inner electron is more tightly bound to the nucleus (the energy of the system is more negative) than an outer shell electron; therefore, more energy is required to remove it. When an inner electron is removed, an electron in a higher energy level drops down to take its place. The radiation emitted is generally higher than ultraviolet, and these energies on the electromagnetic spectrum are called *x-rays.* X-rays are relatively high-energy electromagnetic radiations that are produced by electron transitions from higher to lower energy levels, particularly in more complex atoms. X-rays are also produced when high-speed electrons are stopped in a dense material. This method of x-ray production is consistent with Maxwell's theory that an accelerated electric charge radiates electromagnetic energy.

A complete model of the structure of a multi-electron atom is very complex. The spectral analysis of the energy levels and sub-levels in an atom is simplified somewhat by the fact that in general it is only electrons in the last shell that are involved in the emission of the radiation that is being studied. Although there are some gaps in our understanding of the structure of complex atoms, the overall success of the quantum theory in producing a model of the atom that is generally consistent with experimental chemistry and physics marks this as one of man's most significant steps toward understanding physical reality.

LASERS

The model of the atom that we have developed has done more than just provide a theoretical model that is consistent with observations. It has also led to the discovery of practical devices that benefit man. One of the most significant applications of our understanding of atomic structure is the laser, which was built in the 1950's. Laser is an acronym for Light Amplification by Stimulated Emission of Radiation. Actually, the laser was preceded by the maser (Microwave Amplification by Stimulated Emission of Radiation), but since the laser has much wider applications, we shall restrict our consideration to it.

If we analyze the interactions between radiation and the electrons in an atom, we find that there are three possible interactions: absorption, spontaneous emission, and stimulated emission. The absorption process is one in which an electron is raised to a higher energy level by absorbing electromagnetic energy. Spontaneous emission of radiation occurs when an electron that has been raised to a higher energy level returns to a lower level. It is only the stimulated emission of radiation by atoms that we have not considered, and it is this process that is involved in the laser.

Atoms in an excited state can be stimulated to emit their excess energy by subjecting them to radiation. In order for such emission to occur, the energy of a photon of the stimulating radiation must equal the amount of energy the electron has above its unexcited level. The reason this interaction is not observed in a normal collection of excited atoms is that the ratio of atoms in the ground state to atoms in the excited state is very large. Calculations show that an electron remains in an excited state only on the order of 10^{-6} second. Therefore, the net effect of radiation falling on such a collection of atoms is absorption, not emission. What we need to do is to cause a *population inversion*, that is, produce a situation in which more atoms in the sample are in an excited state than are in the ground state. A population inversion can be achieved in some atoms through a process called *pumping*. Pumping is done by an electric discharge, by a very

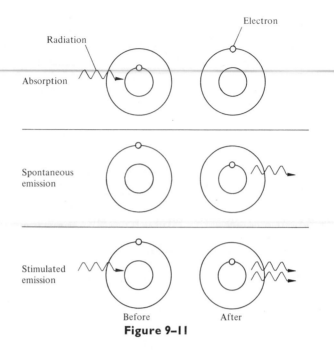

Figure 9–11

intense beam of radiation, or by heat. Atoms that lend themselves to pumping have an excited energy level near another energy level in which electrons can remain for a relatively long time. As shown in Figure 9–12, an electron in the lowest energy level, A, is pumped to the excited level, C. From C, the electron drops to level B almost immediately, but then remains in level B long enough to be stimulated to emit radiation. The energy E of a photon of the stimulating radiation must be such that $E = hf = E_B - E_A$. Essentially, then, we send one photon in and get two photons out. Therefore, light has been amplified in this process. By enclosing the lasing atoms in a chamber with mirrors at either end, the stimulated radiation can be reflected back and forth, stimulating more emissions, and thus tremendous amplification can be achieved. There are two basic types of amplified light: pulsed and continuous.

Amplification of light is not the only significant feature of a laser. The photons emitted all have the same frequency, are traveling in the same direction, and are in phase with each other. A laser thus produces coherent light. Since light from ordinary sources is incoherent, the laser is very useful as a source of coherent light. (The significance of coherence was

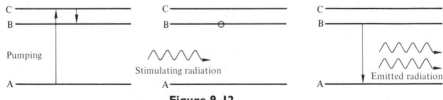

Figure 9–12

discussed previously in connection with waves.) Light from a laser is amplified, monochromatic, parallel, and coherent.

Probably never in the history of science and technology has a discovery found as many significant, diverse applications in as short a time as has the laser. A pulsed laser can burn a hole in a razor blade. The power of the laser, along with its highly directional characteristics, makes it an excellent tool for welding microscopic wires in miniature electronic components. A similar technique on a lower energy scale is being used routinely for welding detached retinas in the eye, thus eliminating very difficult surgery. Although it is still in the experimental stage, the use of the laser in many types of surgery now seems feasible. One great advantage of the laser over the scalpel is that the laser cauterizes as it cuts, making surgery practically bloodless.

Another quite different category of applications of the directional characteristics of the laser is its use in alignment and distance determination. The use of the laser to align everything from tiny research components in a laboratory to tunnel drilling equipment is commonplace. The use of a reflected laser beam to determine the earth-moon distance to within 15 centimeters is well known. In chemical research, the laser is pinpointing chemical reactions on a molecular level. A very interesting feature of laser light that holds great promise for the future is its ability to produce a three-dimensional image, called a *hologram*, on two-dimensional photographic film. Three-dimensional television may not be very far away.

Of all the applications of the laser, the one with the greatest potential probably is its use in communications. The ability of a laser beam (because it is narrow and coherent) to carry much more information than any existing communication system will soon bring needed relief to already overcrowded communication facilities. The laser, however, is only an infant—its most important applications may still be awaiting a discoverer.

EXERCISES

1. Give some examples from your everyday experience of quantized things.

2. These experimental data resulted from an investigation of the photoelectric effect. Set up a rectangular coordinate system so that the horizontal axis is near the middle of the graph paper and you can label the ordinate with both positive and negative values. Plot the maximum kinetic energy of electrons as a function of the frequency of the light.

Frequency of Light (Hertz)	Maximum Kinetic Energy of Electrons (Joules)
6.8×10^{14}	1.8×10^{-19}
6.1×10^{14}	1.3×10^{-19}
5.5×10^{14}	0.9×10^{-19}
5.2×10^{14}	0.8×10^{-19}

(a) From the resulting curve, determine the value of Planck's constant. (b) Extrapolate the curve until it intersects the vertical axis. What is the significance of the fact that this intercept is a negative energy value?

3. The frequency of a certain FM radio station carrier wave is 10^8 hertz. What is the energy of a quantum of this radiation?

4. The wavelength of certain ultraviolet radiation is 2×10^{-7} meter. What is the energy of a photon of this radiation?

5. The cooking elements on top of an electric range typically glow a dull red when they are heating. (a) At approximately what wavelength is the visible energy radiation a maximum? (b) Approximately what is the energy of a single photon of this energy?

6. (a) What is the wavelength of an electron (mass $= 9.1 \times 10^{-31}$ kg) moving at one-half the speed of light? (b) What wavelength would you have if you were cruising along at 60 miles per hour (134 m/sec)?

7. Suppose that you decided to draw to scale a diagram of an atom. You represent the nucleus with a circle of 1 cm radius. Approximately how far away will you draw the boundary of the atom?

8. (a) Determine from the diagram of representative spectra the approximate wavelengths of the four visible lines in the hydrogen spectrum. (b) Verify these wavelengths by substituting $n = 3, 4, 5$, and 6 in the Balmer formula.

9. (a) How much farther away from the nucleus is the tenth Bohr orbit in a hydrogen atom than the first orbit? (b) How much more energy is required to raise the electron from the first to the tenth energy level than from the first to the second?

10. Figure 9–13 shows an energy level diagram for a hypothetical atom. The transition from $n = 4$ to $n = 2$ results in the emission of yellow light. Determine for each of the following transitions what color results or whether it is infrared or ultraviolet: (a) $n = 2$ to $n = 1$. (b) $n = 3$ to $n = 2$. (c) $n = 5$ to $n = 2$. (d) $n = 5$ to $n = 4$. (e) $n = 5$ to $n = 1$.

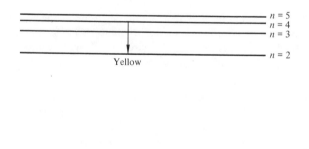
Yellow

Figure 9–13

11. Suppose that the electron in the atom in Exercise 10 is in the $n = 5$ state. (a) By what possible routes can it return to ground state? (b) How many spectral lines will be emitted for each route?

12. (a) Since $L = 2\pi r_n/n$, what is the wavelength of an electron in the second energy level in a hydrogen atom? (b) Using $L = h/mv$, find the speed of this electron ($m = 9.1 \times 10^{-31}$ kg).

13. Excited hydrogen gas produces a spectrum consisting of four visible lines and numerous infrared and ultraviolet lines. Explain how hydrogen, with only one electron, can produce so many lines.

14. Represent in a diagram the configuration of electron orbits in the following atoms: (a) Lithium. (b) Oxygen. (c) Sodium. (d) Argon.

15. Why do fluorine and chlorine have very similar chemical and physical properties (consider the electron configurations)?

16. Explain the existence of the elements between calcium and gallium, in the fourth period of the periodic table, which do not have elements in the first, second, and third periods under which to be placed.

17. The elements iron (26 electrons), cobalt (27 electrons), and nickel (28 electrons) have similar properties because each has two electrons in its outside shell. Show how this is possible by writing the electron configuration for each of these elements.

18. As an aid to visualization, we sometimes picture the inside of an atom as a miniature solar system with the nucleus as the sun and the electrons as the planets. What fundamental differences exist between the solar system and an atom, and how may they cause this analogy to be misleading?

19. Want to exercise your imagination a little? Devise a possible model of the atom that is different from the Bohr model yet is consistent with all observations of atomic phenomena that we have considered in this chapter.

20. (a) Explain how a laser is used to determine the earth-moon distance. What characteristics of the laser beam make this experiment possible? (b) A beam of light from a laser is directed on a diffraction grating. What do you expect to observe on a screen placed beyond the diffraction grating?

10

THE NUCLEUS AND
ITS ENERGY

The model that we have developed thus far represents the atom as being somewhat like a mini solar system. The model has a very small nucleus that contains all of the positive charge and most of the mass of the atom. Around this nucleus is a configuration of electrons. Each electron is in some definite energy state, and energy is radiated from the atom when an electron changes from a higher to a lower energy level. The arrangement of the electrons in the energy levels determines the "chemistry" of the atom. It is clear at this stage that we have a fairly complete model of the atom. But do we? What is the composition and internal structure of the nucleus? To answer this question, we must either study some energy that comes from inside the nucleus or investigate what happens to some energy that we send into it. First, let us look for energy that originates in the nucleus.

RADIOACTIVITY

In 1896, the French physicist Becquerel was investigating the relationship between x-rays and fluorescence. One of the materials that he happened to be studying was a compound of uranium. Some of this uranium salt was placed in a drawer with unexposed photographic plates. Upon removing the plates, Becquerel discovered that they were exposed, even though they had been wrapped in light-tight black paper. On the basis of some further experimentation, Becquerel suggested that the uranium was emitting energy that, after penetrating several layers of paper, was still capable of exposing photographic plates. He referred to this energy as "radiations actives." In 1898, Marie Curie turned her attention to this new phenomenon and coined the word "radioactivity"

242

to describe this form of energy. She and her husband discovered two new elements that exhibited this activity—radium and polonium. By 1904, some 20 radioactive elements were known. Although many scientists were involved in the development of our understanding of radioactivity, the most significant contributions during the first third of the nineteenth century were made by Rutherford and his associates and students.

These early experimenters discovered that radioactivity has some interesting properties. It exposes photographic film, ionizes gases, produces scintillations (flashes of light) in certain materials, penetrates matter, kills living tissue, releases large amounts of energy with little loss of mass, and is not affected by chemical and physical changes of the radiating material. It is the last of these characteristics that is particularly significant to us in our model building. If radioactivity is assumed to originate within the atom, yet is not affected by chemical changes, then it must not be associated with the electrons that are involved in chemical reactions. This suggests that radioactivity comes from the nucleus, and that we might obtain some information about the nucleus from a study of this energy.

Our analysis of radioactivity begins with a consideration of its nature. Is it a wave or a particle? Is it charged or not? The experiment that most completely reveals the nature of radioactivity is one in which the radiation is directed through an electric field produced by two parallel charged plates (Figure 10–1). The result of this experiment is surprising. The single beam of radiation is split into three beams by the electric field! The deflection toward the negatively charged plate indicates a positively charged beam, the deflection toward the positively charged plate indicates a negatively charged beam, and the undeviated beam has no charge. Since the nature of these three beams was not understood at first, they were identified simply as alpha rays (positive charge), beta rays (negative charge), and gamma rays (zero charge). Therefore, radioactivity consists of three

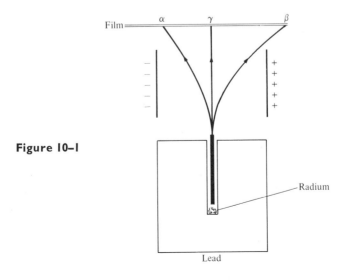

Figure 10–1

types of radiation, and our question regarding the nature of radioactivity thus becomes threefold, since we now must seek the nature of alpha, beta, and gamma radiation.

Experiments revealed right away that gamma rays are the most penetrating and alpha rays are the least penetrating, but the exact nature of each of the three types of radioactivity was not known for several years. On the basis of many experiments by numerous scientists, the conclusion was reached that the natures of these three forms of energy are as shown in Table 10-1.

TABLE 10-1   RADIATION CHARACTERISTICS

	CHARGE	MASS	ENTITY
alpha rays (α)	2 electron charges	7000 electron masses	helium nucleus
beta rays (β)	same as electron charge	same as electron mass	high-speed electron
gamma rays (γ)	none	none	electromagnetic wave

When an unstable atom emits alpha, beta, or gamma radiation, it of course loses energy. This process is called *radioactive decay*. The word "decay" refers to a decrease in energy. The decay of a particular atom is a purely random event. All we can do is assign to a given unstable atom a probability that it will decay in a certain time interval. If it does not decay in that interval, it has the same probability of decaying in the next identical time interval, and so on.

Although little can be said about an individual unstable atom, a great deal can be said in a statistical sense regarding a collection of many unstable atoms. It is a fact that each unstable atom has its own peculiar probability of decay. Therefore, it is meaningful to say that for a large sample, on the average, a given fraction of the atoms will decay in a certain time. The time interval generally used is the time required for half of the unstable atoms to decay. This interval of time is known as the *half-life* of the given collection of atoms. A particular radioactive species always decays at the same rate, and therefore has a unique half-life. Furthermore, the rate of decay is not affected by physical or chemical changes in the sample. Since the activity of a sample is proportional to the number of unstable atoms present, the half-life is also the time in which the activity will decrease to one half its original value. We find experimentally that the greater the activity (or number of unstable atoms remaining), the greater the number that will decay in a given time interval. This law of radioactive decay is shown in Graph 10-1 for any radioactive sample. The quantity that varies from one radioactive species to another is the amount of time represented by the half-life. Half-lives range from fractions of a second for some artificially produced unstable atoms to billions of years for some naturally occurring radioactive elements.*

---

* It is recommended that Experiment 14 in the lab manual be done at this time.

GRAPH 10-1

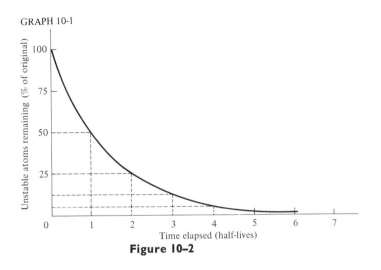

Figure 10–2

Consider two samples, each of which initially contains 100,000 unstable atoms. Sample A has a half-life of one week and Sample B has a half-life of two weeks. The decay curves for these two samples are shown in Graph 10-2. Note that for both samples, 50,000 disintegrations occurred during the first half-life (100,000 present at the beginning of the interval), 25,000 disintegrations occurred during the second half-life (50,000 at the beginning), 12,500 disintegrations occurred during the third half-life (25,000 at the beginning), and so on. Therefore, the number of disintegrations is proportional to the number of unstable atoms present.

When alpha, beta, or gamma energy encounters matter, it interacts with electrons in outer shells of atoms in the matter. Some of the energy is given to an electron and either raises it to a higher energy level (excites) or, as is more likely, removes the electron from the atom (ionizes). After

GRAPH 10-2

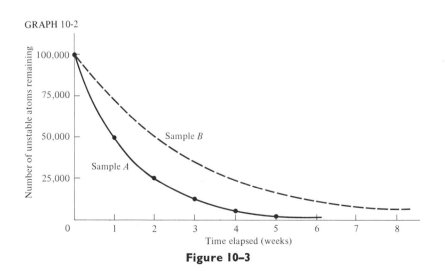

Figure 10–3

each such interaction, the radioactive radiation either disappears or goes on its way with reduced energy. This loss of radioactive energy to matter is absorption. If enough excitations and ionizations are produced, the radiation is completely absorbed. It is interesting to note that alpha, beta, and gamma rays do not make other matter radioactive—they excite and ionize and thus affect the extranuclear atom only.

Since radioactivity cannot be turned off or accelerated, the only control man has of it is through the absorption process. The only way to turn off a radioactive sample is to absorb the energy it is producing. Alpha particles are completely absorbed by a sheet of paper. Beta particles require a greater thickness of more dense matter, such as aluminum, for effective absorption. Gamma radiation is so penetrating that several centimeters of lead are required for effective absorption.

In order to investigate radioactivity, some method of detecting it must be found. Since none of our senses responds to radioactivity, we turn to the effects radioactivity has on matter. Radioactivity is detectable directly by its exposure of film, its production of scintillations, and its ionization of gases. The most practical and widely used of these methods is the ionization of gas atoms. This is the method employed in the Geiger counter. The Geiger counter gives a measure of radioactivity by counting the number of ionizations produced by it. The counter is scaled to read ionization events per unit time, usually in counts per minute (cpm). The unit of activity (rate of disintegration) of a radioactive source is the *curie* (Ci). One curie is approximately the disintegration rate of 1 gram of radium-226 and is defined as exactly 3.7×10^{10} disintegrations per second.

The most exciting idea resulting from a study of radioactivity is the fact that the nucleus appears to be a storehouse of vast amounts of energy. This energy is in the form of the alpha, beta, and gamma radiations that exhibit their energy in the ionizations they produce. Furthermore, for a large collection of atoms, the storehouse seems to be almost inexhaustible. For example, 1 gram of radium liberates approximately 580 joules/hour through radioactive disintegration. After 1600 years, this original 1 gram sample will still liberate approximately 290 joules/hour!

Radioactivity is a naturally occurring phenomenon. However, as we will see later when we discuss nuclear transmutations, stable atoms can be made radioactive by bombarding them with high-energy particles. Most of the radioactive materials used in medicine, industry, research, and other fields are artificially produced from stable atoms.

Man has found many uses for energy released from the nucleus in the form of radioactivity. The high predictability of the statistical rate of decay of a given radioactive sample has proven to be an excellent method of determining the ages of artifacts that are thousands of years old and of rocks that are millions and even billions of years old. The fact that extremely small quantities of radioactive materials can be detected with a Geiger counter makes these radioactive atoms very useful as tracers in disease diagnosis and metabolism studies in human beings and in fertilizer uptake analysis in plants. The penetrating ability of this radiation is applied in thickness gauging in industry. The ionizing effect of radioactivity is the most effective non-surgical method known for cancer therapy. This same ionization can be used in another way to preserve food by killing bacteria. Many research stations in remote areas on Earth and the scientific experiments left by the astronauts on the moon are powered by electricity that is produced by heat resulting from the absorption of radioactivity in matter.

It is clear that many useful applications have been made of this energy from the nucleus, but what does it tell us about the nucleus itself? Some hint is given about the composition of the nucleus from the fact that helium nuclei and electrons emerge from it. However, the more significant information about the nucleus obtained from a study of radioactivity relates to the structure of the nucleus. This knowledge comes from the fact that the alphas, betas, and gammas associated with the nucleus of a given atom have specific energies. The typical radionuclide emits either an alpha or a beta particle followed by gamma radiation. This gamma radiation is an electromagnetic wave that on close examination is found to exhibit a spectrum of frequencies similar to the bright-line light spectra produced by electron energy changes in the atom. A given radionuclide always produces the same gamma spectrum; different nuclei produce different spectra. Thus, the gamma spectrum is the "fingerprint" of a nucleus in the way that a light spectrum is the fingerprint of an atom. Therefore, we conclude from this analogy that there are quantized energy levels in the nucleus. But what occupies these energy levels in the nucleus? What in the nucleus corresponds to the electrons in the extranuclear atom? Of what is the nucleus composed?

NUCLEAR COMPOSITION

The nucleus contains most of the mass of the atom and positive charges equal in number to the orbital electrons. We have agreed that the elements can be built up, as far as their chemistry is concerned, by starting with hydrogen and adding one electron at a time to the vacant energy levels. In order to hold these electrons, positive charges must be added one at a time to the nucleus. Also, it can be determined experimentally that as the number of charges in an atom increases, the mass of the atom increases. An idea that immediately catches our attention is the possibility that all atoms are composed of hydrogen atoms. If we remove the single electron from hydrogen (ionize the atom), there remains a hydrogen nucleus with one unit of positive charge. Since this nucleus has a smallest unit of charge, it is reasonable to assume that it is a fundamental unit of matter and to give it a name—proton. The proton is found to have some 1800 times the mass of an electron. With this information in mind, let us test our hypothesis and build a helium atom from two hydrogen atoms.

For helium, we have two electrons and a nucleus with two protons, and thus a charge of +2—so far, so good. Now let us check the mass of the helium atom. Here we run into a stumbling block, for the mass of the helium atom is approximately four times the mass of the hydrogen atom!

Since the mass of the electron contributes little to either atom, the mass discrepancy must be in the nucleus. A similar mass discrepancy occurs in all other atoms also. This suggests the existence of something in the nucleus that contributes mass without adding charge.

What is the form of the extra mass in the helium atom? At this stage in our development, the only particles we have to work with are electrons and protons, both of which are charged. But wait! Suppose that the helium nucleus consists of four protons and two electrons—net result, +2 charge and four times the mass of hydrogen. That is certainly a reasonable model of the nucleus, but theoreticians tell us that it is impossible. We can show by calculations based on sound physical principles that it is not possible for an electron with its low mass and resulting high speed to be contained in a space as small as a nucleus. Another good idea shot down.

What is your idea? One solution, of course, is simply to assume the existence of a particle that has no charge and a mass approximately equal to the proton's mass. That is exactly what was done for several years; the existence of such a particle was assumed until it was discovered experimentally in the early 1930's. This particle with no charge and with mass slightly greater than the mass of a proton is called a *neutron*. Therefore, the model that has resulted in no inconsistencies says that a helium atom consists of two electrons, two protons, and two neutrons. Protons and neutrons, which are the primary constituents of the nucleus, are collectively referred to as *nucleons*.

We will now find it convenient to denote the number of electrons and also the number of protons in an atom with a quantity called the *atomic number*, represented by the symbol Z. The number of nucleons is called the *mass number* and is denoted by A. These two numbers can be used in conjunction with the chemical symbol for the element (say, X) to indicate the constituents of an atom completely in the following way: $^A_Z X$. Thus, helium, with an atomic number 2 and mass number 4, is written $^4_2$He. The number of neutrons in a nucleus is the difference between the mass number and the atomic number. Symbolic representations of some common particles are given in Table 10–2.

An element, and therefore a particular atom, is determined by the number of electrons (Z). This atom must also have Z protons. However, since the neutron has no charge, it does not influence the number of

TABLE 10-2.

Particle	Symbol
Alpha	$^4_2$He
Beta, electron	$^0_{-1}$e
Positron	$^0_1$e
Gamma	γ
Proton	$^1_1$H
Deuteron	$^2_1$H
Neutron	$^1_0$n

electrons in the atom. Thus, although two atoms of the same element must have the same number of protons, they can have different numbers of neutrons. Atoms that have the same number of protons (same atomic number) but different numbers of neutrons (different mass numbers) are called *isotopes* of the given element. If such an atom is radioactive, it is said to be a radioisotope of the element. For example, hydrogen has three isotopes: ^1_1H ("normal" hydrogen), ^2_1H (deuterium), and ^3_1H (tritium). Tritium is unstable and therefore is a radioisotope of hydrogen (Figure 10–4). An isotope of an element is frequently indicated by giving the chemical symbol followed by the mass number of the particular isotope; H-2, Co-60, U-238.

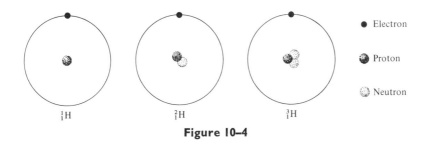

^1_1H ^2_1H ^3_1H ● Electron   Proton   Neutron

Figure 10–4

The so-called "atomic weight" of an element that appears in a periodic table of the elements is an average of the masses for the naturally occurring isotopes of the element. Since the value given is a weighted mean, the mass number for the most abundant isotope of an element is assumed to be the integer nearest to the value given in the periodic table. For example, in the periodic table chlorine has an atomic number of 17 and a mass of 35.45. Thus, a chlorine atom contains 17 electrons and 35 nucleons (17 protons and 18 neutrons).

NUCLEAR TRANSMUTATIONS

When a radioisotope of an element emits either an alpha particle or a beta particle, the charge of its nucleus changes. This modified charge of the nucleus causes either a loss or a gain of orbital electrons, and therefore an atom of a different element is the result. Thus, along with the energy decay associated with alpha or beta particle emission, there is also the transmutation of the unstable atom into a different species of atom.

How do we analyze such a transmutation? When an alpha particle leaves a nucleus, it takes with it (in addition to energy) two units of positive charge and four units of mass (assuming that the mass of a proton = the mass of a neutron = a unit mass). When a beta particle exits, it removes one unit of negative charge and negligible mass ($\sim$1/1800 unit mass) from the nucleus. Gamma radiation, of course, removes only energy. By applying two of the most basic principles of physics—the conservation of charge and the conservation of mass—a rule for determining the results of alpha and beta decay can be formulated: In alpha decay, the new atom has lost two charges and four mass units; in beta decay the new atom has gained one charge and has experienced no significant mass change. In both alpha and beta decay, the new nucleus is usually left in an excited state and thus emits gamma radiation in order to get rid of the excess energy. There is also the possibility that the nucleus produced by the original decay will be another radioisotope of some element that again emits an alpha or beta particle.

A convenient way to apply the conservation principles to radioactive transmutations is demonstrated in the following examples (an asterisk following the chemical symbol indicates that the nucleus is in an excited state):

$$^{238}_{92}U \rightarrow {}^{4}_{2}He + {}^{234}_{90}Th* \qquad {}^{234}_{90}Th* \rightarrow {}^{234}_{90}Th + \gamma$$

$$^{131}_{53}I \rightarrow {}^{0}_{-1}e + {}^{131}_{54}Xe* \qquad {}^{131}_{54}Xe* \rightarrow {}^{131}_{54}Xe + \gamma$$

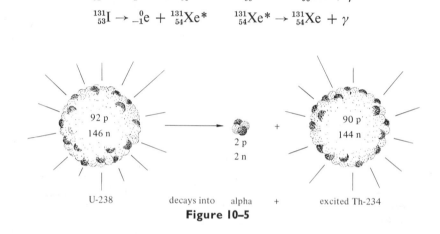

U-238   decays into   alpha   +   excited Th-234

Figure 10–5

These transmutations are natural in that they are the results of spontaneous emissions from unstable nuclei. It is also possible to induce transmutations in stable nuclei by bombarding them with subatomic particles. The first such alchemist's-dream-come-true occurred in 1919 in the transformation of nitrogen into oxygen according to the reaction:

$$^{14}_{7}N + {}^{4}_{2}He \rightarrow {}^{1}_{1}H + {}^{17}_{8}O.$$

The isotope of oxygen produced from nitrogen in this reaction is a stable atom. However, many such induced reactions result in the production of radioisotopes that then decay to isotopes of other elements. These man-made radioisotopes emit alpha particles, beta particles, gamma radiation, and positrons (beta particles with a positive charge). For example:

$$^{27}_{13}Al + {}^{4}_{2}He \rightarrow {}^{1}_{0}n + {}^{30}_{15}P \qquad {}^{30}_{15}P \rightarrow {}^{30}_{14}Si + {}^{0}_{1}e.$$

Thus, stable nuclei are made radioactive by bombardment with particles.

In order to induce nuclear transmutations, particles must have sufficient energy to penetrate, or at least get very close to, atomic nuclei. The naturally occurring particles with this much energy are limited essentially to alpha and beta particles. Even for these, the energy range is restricted. The need for greater varieties of particles and energies for bombarding nuclei has led to the development of *particle accelerators*. A particle accelerator is a device for producing beams of subatomic particles of various energies. Studies of the interactions of these energetic particles with nuclei have added much to our knowledge of the structure and properties of the nucleus. In order to probe the mysteries of the nucleus and to produce radioisotopes, particle accelerators have therefore been designed to increase the energy of subatomic particles and to direct them to a specific target. Typical bombarding particles are alphas, protons, deuterons, and neutrons. Usually the transmutations induced by these particles involve the emission of one of the four particles.

Although there are many types of accelerators, most of them are basically either circular or linear in design. In either type, the acceleration of charged particles is accomplished by electric or magnetic fields or both. The principle of operation of two of the most important circular accelerators, the cyclotron and the synchrotron, was discussed in Chapter 5. The simplest linear accelerators (for example, the Van de Graaff and the Cockcroft-Walton accelerators) increase the energy of charged particles by creating a large potential difference between two (or more) electrodes. The particles are thus accelerated by an electric field. In a more sophisticated linear accelerator (LINAC), the particles are accelerated by radio frequency waves.

The neutron, because it is electrically neutral, cannot be accelerated by any of these devices. Neutrons of very high energy are obtained by bombarding certain materials with other high-energy particles, such as protons and deuterons. These ejected high-energy neutrons can then be passed through materials called *moderators*, and the resulting collisions slow them to any energy desired.

The particular effectiveness of neutrons in inducing radioactivity has led to an entire area of investigation called neutron activation analysis. An example of neutron activation is:

$$^{27}_{13}\text{Al} + ^{1}_{0}\text{n} \rightarrow ^{1}_{1}\text{H} + ^{27}_{12}\text{Mg}^* \qquad ^{27}_{12}\text{Mg}^* \rightarrow ^{27}_{12}\text{Mg} + \gamma.$$

Neutron activation of the heaviest isotope of an element found in nature, U-238, is particularly interesting. This reaction results in U-239, which decays by beta emission and thus goes up one unit in charge. But uranium is the natural element with the highest charge. Therefore, this reaction has produced a new element called neptunium. Np-239 is also unstable by beta decay and produces an element one charge higher still, known as plutonium:

$$^{238}_{92}\text{U} + ^{1}_{0}\text{n} \rightarrow ^{239}_{92}\text{U} \qquad ^{239}_{92}\text{U} \rightarrow ^{0}_{-1}\text{e} + ^{239}_{93}\text{Np}$$

$$^{239}_{93}\text{Np} \rightarrow ^{0}_{-1}\text{e} + ^{239}_{94}\text{Pu}.$$

Man has produced a total of at least 12 such transuranic elements through induced transmutations.

It appears that the maximum number of neutrons and protons that can be contained in a particular nucleus is determined by some physical law that is not yet completely understood. If a nucleus contains an excess of either nucleon, it is unstable and gets rid of the excess by emitting a particle. The theory of beta decay is particularly interesting. Have you wondered how an electron (a beta particle) can be emitted from a nucleus when it is theoretically impossible for a free electron to be contained in the nucleus?

The answer is that a neutron can change into a proton and an electron. Thus, in beta decay a neutron in the nucleus is converted into a proton plus an electron; the electron, since it cannot exist in the nucleus, is emitted as a beta particle. The net result of beta decay as far as the nucleus is concerned is that it has lost a neutron and gained a proton. In decay by positive beta (positron) emission, a proton is converted into a neutron with the creation, in order to conserve charge, of a positron that is immediately emitted.

FISSION

You recall that when uranium-238 is bombarded with neutrons it becomes U-239, which decays by beta emission to neptunium, which in turn emits a beta particle to transmute to plutonium. A small amount (less than 1 per cent) of naturally occurring uranium consists of another isotope, U-235. Strange things happen when this isotope of uranium is bombarded with neutrons. Instead of producing a heavier atom, neutrons cause U-235 to split (called fission) into two atoms of smaller masses. This fission reaction is made even more interesting by the fact that a large amount of energy is released in the process. But that is not all. What is even more exciting is the fact that in this reaction, two or three neutrons are released (Figure 10–6). In other words, some of the same "bullets" that produced the reaction are available to cause similar reactions. A typical fission reaction for U-235 is:

$$^{235}_{92}U + ^{1}_{0}n \rightarrow ^{144}_{56}Ba + ^{89}_{36}Kr + 3\,^{1}_{0}n + \text{energy}.$$

Both the barium and the krypton produced in the reaction are radioactive. The energy released per fission is on the order of one million times the energy involved in a chemical reaction!

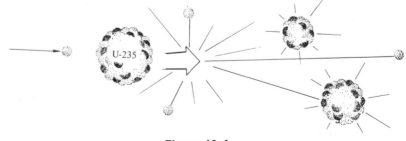

Figure 10–6

Thus we have in the fission reaction not only a source of tremendous quantities of energy, but also an energy source that is capable of sustaining itself once it is triggered by a neutron. In order to have a sustained reaction, at least one of the neutrons produced in the fission must produce another fission. The problem is that the uranium that comes from the Earth's surface contains only about one atom of U-235 for every 140 atoms of U-238. Therefore, the probability is good that a given neutron will encounter a U-238 nucleus and produce U-239 rather than cause a fission reaction. Thus, in order to have a sustained reaction, we must increase the ratio of U-235 atoms to U-238 atoms. This "enriching" of uranium is not easy, since two isotopes of the same element cannot be separated chemically. Two techniques that have been employed successfully to separate U-235 from U-238 are gaseous diffusion and electromagnetism. Both methods use the mass differences of the isotope for the separation.

There is another problem in maintaining the fission reaction. Some of the neutrons will escape from the uranium without producing a reaction. If too many escape, the reaction will not be sustained. The way to increase the probability of an interaction is simply to increase the size, or mass, of the sample. Thus, there is a "critical mass" of U-235, below which the reaction will not sustain itself. Therefore, if we have a sufficient ratio of U-235 to U-238 and the mass of the sample is large enough, then a chain reaction results. Each time a neutron reacts with a U-235 nucleus to cause fission, energy is released (Figure 10–7).

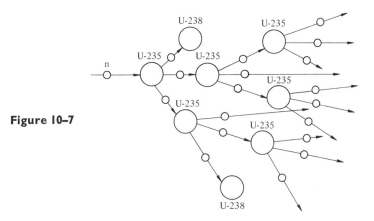

Figure 10–7

In order to increase the efficiency of nuclear fission, one thing further can be done: to slow, or moderate, the neutrons. A slowly moving neutron is more likely than a faster neutron to be captured and to produce a fission reaction. Neutrons are slowed by collisions with certain materials. The most widely used moderators for neutrons are graphite (carbon) and heavy water (H_2O molecules in which the hydrogen atoms have one neutron and are, therefore, deuterium). Fission reactions are controlled (turned on and off and accelerated) by absorbing neutrons in a control rod made of material such as cadmium, which readily absorbs neutrons.

As we mentioned previously, the separation of U-235 from U-238 presents quite a problem. Fortunately, there is another way to obtain fissionable material. The plutonium-239 produced by neutron bombardment of U-238 also fissions when exposed to a neutron flux. Since Pu-239 is a different element, it can be separated chemically from uranium.

Another interesting fact about Pu-239 is that it is produced as a by-product from the U-238 that is always present in a U-235 reactor. This concept has spurred the development of *breeder reactors*, reactors that use some of the fission neutrons to produce more fissionable fuel in the form of Pu-239. Breeder reactors can actually produce more fissionable fuel than they consume. The fission reaction for Pu-239 is similar to that for U-235 in that it releases a tremendous amount of energy and frees neutrons in the process. Unfortunately, its waste products are also radioactive. The energy released from the fission of 1 kilogram of plutonium is equivalent to the chemical energy released from some 50 railroad cars of coal.

Many nuclear reactors all around the world are now heating water to generate electricity. As supplies of fossil fuels diminish, nuclear reactors will assume an ever-increasing role in the production of electrical power.

What is the source of the fantastic energy released in the fission process? The answer is found in the difference in total mass of the particles before the reaction and the total mass after the reaction. The fact is that the sum of the masses of the products is less than the sum of the masses of the reactants; that is, mass has disappeared in the reaction! One of the conclusions of Einstein's special theory of relativity (discussed in Chapter 8) is that mass and energy are really different manifestations of the same entity. Thus, in the fission reactions, mass m has been converted to an amount of energy mc^2. Most of this energy appears as kinetic energy of the fission fragments.

The masses involved in the subatomic realm are of course extremely small. For this reason, it is convenient to define a unit of mass appropriate to this level. Such a standard is the *atomic mass unit* (amu), which is defined as exactly one twelfth the mass of the carbon-12 isotope. One amu is equivalent to 1.66×10^{-27} kg. The mass of a proton is approximately 1.007 amu and the mass of the neutron is slightly greater. The energy equivalent of one amu is 1.49×10^{-10} joules. The magnitude of this quantity means that the electron volt (defined in Chapter 9) is a more practical energy unit than the joule. The energy equivalent of one amu is 931×10^6 eV, or 931 MeV (million electron volts). A chemical reaction involves several electron volts of energy, whereas a fission reaction typically releases several million electron volts of energy.

FUSION

Fission is not the only nuclear reaction in which a mass-to-energy conversion occurs. Interestingly enough, mass is converted to energy in the combination, or fusion, of two lighter nuclei to form a heavier nucleus.

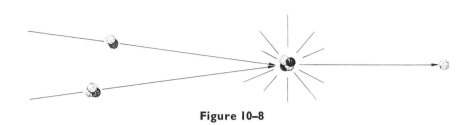

Figure 10–8

To get an idea of what is involved in such a reaction, suppose we consider putting two protons and two neutrons together to form a helium nucleus (Figure 10–9):

mass of 2 protons	2.0146 amu
mass of 2 neutrons	2.0173 amu
total mass of 2 p and 2 n	4.0319 amu
mass of helium nucleus	4.0015 amu
mass lost	0.0304 amu

Figure 10–9

In this process, 0.0304 amu of mass has been converted to some 28 MeV of energy. This is just the energy needed to separate a helium nucleus into its constituents, and therefore is the energy that holds this nucleus together. The energy required to separate the nucleons of any nucleus is referred to as its *binding energy*. The binding energy is associated with the nuclear force that opposes the electrostatic repulsive force between protons, but the exact relationship is not completely understood.

Actually, the liberation of vast amounts of energy via the fusion reaction was discovered through studies of the sun. The perplexing problem concerning the source of energy in the sun and other stars was solved when it was realized that in the sun the conditions most conducive to a fusion reaction are present—very high temperatures and plenty of hydrogen. High temperatures (on the order of millions of degrees Celsius) are required to give the nuclei enough energy to overcome the electrostatic repulsion of the positive charges they contain. Hydrogen nuclei should be the easiest of all nuclei to fuse, since they contain the smallest positive charge. Although there are probably many types of fusion reactions occurring in the sun, the reaction that produces most of the sun's energy is the proton-proton cycle:

$$_1^1H + {_1^1}H \rightarrow {_1^2}H + {_1^0}e$$

$$_1^2H + {_1^1}H \rightarrow {_2^3}He$$

$$_2^3He + {_2^3}He \rightarrow {_2^4}He + 2\,{_1^1}H.$$

In this series of fusion reactions, six protons produce one alpha particle and two protons. Thus the net result is that four hydrogen nuclei have combined to produce one helium nucleus. In this process, mass m is converted into energy mc^2. The amount of energy involved is approximately 25 MeV per cycle.

Since high temperatures (millions of degrees C) are required to initiate a fusion reaction, these are called *thermonuclear reactions*. The attainment of these very high temperatures has been the chief problem in producing fusion reactions on earth. The simplest way of obtaining high enough temperatures (or high enough speeds to drive nuclei together) is through a fission reaction. A fission reaction is used to force hydrogen nuclei together with enough force to cause them to fuse in the thermonuclear (hydrogen) bomb. The only other method devised so far to cause fusion uses plasmas.*

A plasma is an ionized gas. It consists of free positive ions and free electrons, but the collection as a whole is neutral. A plasma can be produced by an electrical discharge, and its temperature can be raised by passing current through it and by compressing it with magnetic fields. For temperatures attainable in plasmas produced on Earth, the proton-proton cycle is very improbable. The carbon cycle, the chief fusion reaction sequence of some stars, takes too long at plasma temperatures to be practical. Therefore, we must look for other fusion reactions as possible sources of controlled thermonuclear energy.

The most probable fusion reaction at attainable plasma temperatures is:

$$_1^2H + {_1^3}H \rightarrow {_2^4}He + {_0^1}n + \sim 18 \text{ MeV energy.}$$

In this reaction, a deuteron and a triton are fused into a helium nucleus. This is probably the reaction used in nuclear weapons. It is not practical as a large-scale energy source because the production of $_1^3H$ is too expensive. On the other hand, deuterium is relatively inexpensive, occurring naturally in water. Although the ratio of deuterium to hydrogen in water is very small, it is fairly easy to separate, and of course, the supply is

---

* The feasibility of using a powerful laser pulse to initiate fusion is currently being investigated.

almost unlimited. Two good prospects for controlled fusion reactions are:

$$\mathrm{_1^2H} + \mathrm{_1^2H} \rightarrow \mathrm{_2^3He} + \mathrm{_0^1n} + 3.2 \text{ MeV energy} \quad and$$

$$\mathrm{_1^2H} + \mathrm{_1^2H} \rightarrow \mathrm{_1^3H} + \mathrm{_1^1H} + 4.0 \text{ MeV energy.}$$

Since $\mathrm{_1^3H}$ is produced in the second reaction, it can now react with $\mathrm{_1^2H}$ to produce $\mathrm{_2^4He}$, a neutron, and 18 MeV energy.

Theoretically, there is no reason that the fusion reaction cannot be used by man as a source of energy. However, formidable obstacles still stand in the way of the utilization of controlled fusion energy. In addition to the problem of maintaining high enough temperatures for any reasonable length of time, there is the very serious problem of containment. In what do we contain a plasma that has a temperature of several million degrees celsius? All known materials disintegrate instantly at such temperatures. Most of the research in containment of the fusion reaction is in the direction of using magnetic fields. This so-called magnetic bottle would suspend the plasma in a vacuum, thus avoiding contact with any material container.

The advantages of the fusion reaction over the fission reaction as a source of energy are twofold. In the first place, the supply of fusion fuel is almost unlimited, but fission fuel is limited by the amount of uranium available in the earth's surface. One gallon of ordinary water contains enough deuterium to produce fusion energy equivalent to the combustion energy of 300 gallons of gasoline. It has been estimated that there is enough deuterium in the oceans to produce the world's current power consumption for billions of years! The second significant advantage of fusion energy is that the waste product is ordinary, stable, inert helium, whereas the waste products of fission are radioactive. What a break for the environment! These two factors make the solving of the enormous problems involved in the utilization of the fusion reaction as an energy source seem worthwhile. There is little doubt that the obstacles will be overcome—it appears to be just a matter of time.

IONIZING RADIATION

Radiation is energy spreading out from a source. In our development of physics, we have considered many types of radiation: sound, light, radio, IR, UV, x-ray, alpha, beta, gamma, neutron, and so on. Some of these radiations ionize matter; some do not. Because those that do ionize have special significance in the world today, let us consider the radiations that are capable either directly or indirectly of removing electrons from atoms.

Some ionizing radiations, such as alphas, betas, neutrons, and fission fragments, are high-speed particles; the remainder are electromagnetic waves. High-energy charged particles interact directly with matter, primarily through Coulomb forces, to produce excitations and ionizations. In these interactions, the charged particle loses some energy, and after removing an electron it goes on its way to produce another ionization (or excitation). A charged particle continues this process until its energy is exhausted, at which time it becomes a part of the material and is said to have been absorbed. Since neutrons, x-rays, and gamma rays have no charge, they ionize directly only by collisions. These radiations owe their ionizing ability to the secondary ionizations that result from these encounters. Neutrons are very effective in producing ionizations indirectly through nuclear transmutations and through collisions with charged particles (particularly protons), which subsequently produce ionizations. X-rays and gamma rays interact with matter to release high-energy electrons through the photoelectric effect and the Compton effect. In addition, high-energy gamma radiation can produce an electron-positron pair by conversion of energy to mass through the relation $E = mc^2$. In all these interactions of x-rays and gamma rays with matter, the chief effects are secondary ionizations produced by the energetic electrons.

In its encounter with matter, ionizing radiation ultimately transfers energy to orbital electrons. The kinetic energy of the freed electrons in turn produces heat, and therefore the net effect is that radiation energy has been transformed into heat energy. The energy of radiation is thus absorbed by matter. The ionized atoms recover electrons, and after irradiation the atoms in the matter are essentially the same as they were before they absorbed the radiation. However, while an atom is temporarily ionized, it may lose its place in the molecular structure. If enough molecules in a living cell are affected, the chemical machinery of the cell may be disrupted to such a point that the cell ceases to function. Therefore radiation, through the process of ionization, kills living matter.

Ionizing radiation is detected by its interactions with matter— exposure of films, production of scintillations, and ionization of gases. The basic unit of radiation exposure is the *roentgen* (R), named after the discoverer of x-rays, Wilhelm Roentgen. One roentgen is the amount of *x*- or gamma radiation required to produce 2.6×10^{-4} coulombs of ionization per kilogram of dry air. The roentgen is an inadequate unit for two reasons: It is not applicable to particle radiation, and it is a unit of exposure, not absorption. For these reasons, another unit, the *rad* (*R*adiation *A*bsorbed *D*ose), is more practical. One rad is 0.01 joule of energy absorbed per kilogram of any substance from any ionizing radiation. For *x*- and gamma radiation, the absorbed dose in rads for water and soft tissue is approximately the same as the exposure in roentgens, but this is not true for particle radiation. A third radiation unit, the *rem* (*R*oentgen *E*quivalent *M*an), makes allowance for the fact that some types of radiation are biologically more effective than others. For example, a one rad dose of alphas is much more effective biologically than a one rad dose of gammas. The most commonly used unit of absorbed dose of ionizing radiation is the rad. A comparison of some radiation doses is given in Table 10–3.

The effects ionizing radiation has on man are of two types: genetic, which can be transferred to offspring, and somatic, which affect only the exposed individual. Genetic effects can occur only if the primary repro-

TABLE 10-3.

SOURCE	APPROXIMATE DOSE OR DOSE RATE
Natural background (cosmic rays and natural radioactivity)	0.15 rad/yr
Fallout in U.S.	0.015 rad/yr
Radium wrist watch dial	0.001 rad/hr
Diagnostic x-ray	
Chest	0.1 rad/film
Dental	0.5 rad/film
Single high-level, whole-body doses First evidence of blood changes	25 rads
LD-50. Lethal dose for one-half of exposed individuals. All suffer severe radiation sickness	500 rads

ductive organs are exposed to radiation. Significant damage is done to chromosomes of reproductive cells by ionizing radiation. The chief results of such damage seem to be reduced fertility and increased occurrence of miscarriages and stillbirths. The birth of deformed individuals as a result of radiation exposure is very rare. The real question in connection with genetic effects is whether or not damage remains hidden for many generations and then shows up as mutations later. Not enough time has elapsed since the advent of significant amounts of ionizing radiation to answer this question.

The somatic effects vary greatly with the dose, ranging from minor skin burns to reduction of resistance to disease, from acute sickness to development of cancer or to death. The body recovers to some extent from radiation damage, but it recovers less and less as the accumulated dose increases. One thing is certain—radiation sickness is not contagious. Radiation effects on the body are not all harmful. The use of carefully controlled radiation to destroy cancer cells is well known. Radiation is an indispensable tool in diagnostic techniques in medicine and dentistry.

How can you protect yourself from ionizing radiation? The two chief factors in radiation protection are distance and shielding. The dose rate from a small source falls off as the square of the distance from the source (inverse-square law), as shown in Graph 10–3. If the dose at 1 meter is 1 rad, then at 2 meters it is 1/4 rad, at three meters it is 1/9 rad, and so on. Shielding charged particles is easy, since they interact strongly with matter. A sheet of paper stops most alpha particles, and a sheet of aluminum will do the same for beta particles. Shielding x-rays, gamma rays, and neutrons, however, is a different matter. These radiations are very penetrating, and therefore relatively large thicknesses of dense materials are required for effective shielding. Although lead is the most effective shielding material, sufficient thicknesses of earth, water, and concrete will also do the job and are commonly used if space is not a problem. Graph 10–4 shows the effectiveness of concrete and lead as shielding materials for gamma radiation from cobalt-60.*

* It is recommended that Experiment 15 in the lab manual be done at this time.

GRAPH 10-3

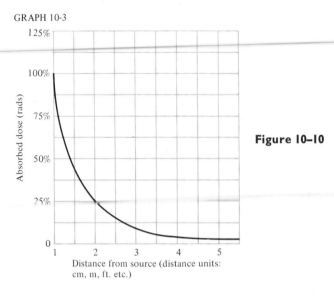

Figure 10–10

Is ionizing radiation good or bad? The answer depends on whether you ask the individual who has had cancer arrested by radiation therapy or the individual who survived the horrors of Hiroshima or Nagasaki. The value of radiation, like that of most technological developments, depends on how man chooses to use it.

GRAPH 10-4

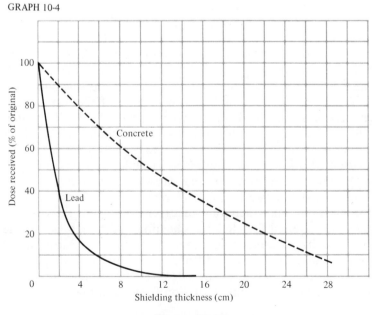

Figure 10–11

ELEMENTARY PARTICLES

Man's unceasing search for the ultimate structure of matter has been spurred through the centuries by his belief that at base, physical reality is simple and knowable. The Greeks' concept of earth, air, water, and fire was replaced by the atoms of the chemical elements around 1800. Early in the twentieth century, atoms yielded the title of elementary particles to the basic constituents of the atom—electrons, protons, and neutrons. With a little explanation, the photon could be considered a fourth particle in the list. At this stage in history, the ultimate structure seemed fairly simple and complete. However, there were many unanswered questions concerning such fundamental concepts as the nature of the force that holds the nucleus together, the apparent non-conservation of energy in beta decay, and the notion of antiparticles. The search for understanding of these basic phenomena in the last few decades has uncovered a host of elementary particles. Let us consider briefly the developments that have produced the current "elementary particle zoo."

Investigations into the nature of the force that holds nucleons together led to the theoretical hypothesis that nucleons are held together by the exchange of an intermediate particle called a *meson* of mass between the mass of an electron and the mass of a proton. This hypothesis was verified by the discovery of the *pi meson*, with a mass of 274 times the mass of an electron. Pi mesons come in three forms—positive, negative, and neutral. Our model of the nature of the nuclear force is that nucleons are held together by exchanging pi mesons. Further investigations revealed that pi mesons, or pions, decay into other intermediate particles called *muons*. Another meson, the *kaon*, was also discovered.

The beta particle that is emitted from an unstable nucleus is found to display a continuous energy distribution from near zero up to some maximum value. The problem is one of accounting for the remainder of the energy in a case in which a beta particle has less than the maximum energy. This problem was solved theoretically by assuming the emission in beta decay of an uncharged, nearly massless particle in addition to the electron. This particle is called a *neutrino*. Because the neutrino does not produce ionizations and interacts only very weakly with nuclei, it proved to be very elusive, and it was first observed more than 20 years after it was postulated theoretically. The neutrino is now assumed to be an uncharged, massless particle.

The discovery of the *positron* (positive electron) was a significant one in two respects—its existence had been predicted theoretically several years before it was observed and it was the first *antiparticle* to be detected. When an electron and positron come together, they annihilate each other and their combined mass is converted into energy through the $E = mc^2$ relation. Other antiparticles were predicted and subsequently discovered. We now know that every particle has its antiparticle. There are antiprotons, antineutrons, antineutrinos, and so on. To be consistent, a photon is considered its own antiparticle.

Recent investigations of very high energy reactions produced by the various particle accelerators have resulted in the discovery of additional "elementary" particles. These particles generally have incredibly short lifetimes and bear such designations as lambda, sigma, and xi particles. The total number of elementary particles now known depends on who is doing the counting. At least 32 (many more by some counts) distinguishable particles have been observed. Others, including *gravitons* (gravity particles), *quarks* (particles that have fractional electric charge), and *tachyons* (particles that go faster than light), have been predicted but not observed. Various schemes for grouping the elementary particles in families have been devised, but there is not complete agreement even on a classification system.

Is there some grand pattern among these particles? Are all these particles really elementary, or are some of them combinations of others? Is there a simple, knowable structure to physical reality? The physicist's motivation for seeking answers to these questions is based on his faith in the basic simplicity and orderliness in nature.

EXERCISES

1. Suppose that you come across a sample of unknown material that a Geiger counter reveals to be radioactive. Describe how you would determine experimentally which type of radioactivity (alpha, beta, or gamma) is being emitted.

2. The data in Table 10–4 are from a study of the radioactive decay of iodine-131. On

TABLE 10–4

DATE	ACTIVITY (cpm)	DATE	ACTIVITY (cpm)
Oct. 26	8920	Nov. 5	3830
Oct. 28	7540	Nov. 7	3110
Oct. 30	6410	Nov. 10	2240
Nov. 1	5340	Nov. 14	1690
Nov. 3	4490	Nov. 16	1520

coordinate paper, plot the activity of I-131 as a function of elapsed time. The time axis should cover at least 30 days. Answer the following questions from the resulting curve: (a) What is the half-life of I-131? (b) How much time is required for three fourths of the atoms to decay? (c) Estimate the activity of this sample on November 25. (d) If this amount of I-131 was originally (on Oct. 26) worth $6.00, what was its value two weeks later?

3. As the result of an accident, some gold-198 (half-life of 2.7 days) is spilled in a laboratory. It is determined that the level of radioactivity is five times greater than that for safe occupancy. How long will it be before people can safely occupy the room?

4. Why is sodium-24 (half-life of 15 hours) a particularly good radioactive tracer to use in the human body?

5. Explain what a "cobalt (Co-60) treatment" does in the treatment of cancer.

6. Explain how an analysis of the carbon-14 (half-life of 5600 years) content of the Dead Sea Scrolls could be used to determine their age.

7. How can radioactivity be used to preserve foods? What about the problem of eating the food after it has been irradiated?

8. Determine the number of electrons, the number of protons, and the number of neutrons for each of the following: ^1_1H, ^2_1H, $^{12}_6C$, $^{16}_8O$, $^{59}_{27}Co$, $^{60}_{27}Co$, $^{131}_{53}I$, $^{235}_{92}U$, $^{238}_{92}U$.

9. Write the following natural transmutations in symbol form and thus find the missing component: (a) Th-232 transmutes by radioactive decay into Ra-228. (b) K-40 decays into Ca-40. (c) C-14 decays by beta emission. (d) Ra-226 emits an alpha particle.

10. Write the following induced transmutations in symbol form and thus find the missing component: (a) When Be-9 is bombarded with alpha particles, a neutron is emitted. (b) When C-12 is bombarded with deuterons, C-13 is produced. (c) When B-10 is bombarded with alpha particles, N-13 is produced. (d) The N-13 just produced is unstable and decays by positron emission. (e) Pu-239 when irradiated with neutrons becomes unstable Pu-241, which then undergoes beta decay.

11. The uranium series, one of four natural radioactive series, begins with U-238 and, through a succession of 14 decays involving the emission of eight alpha particles and six beta particles, terminates in a stable isotope of an element. Determine the element and its isotope number.

12. (a) Calculate the mass of your body in amu. (b) If all this mass could be converted to energy, how much energy in joules would result? (c) How much energy in MeV is represented by the mass of your body?

13. How much mass (in amu) is converted to energy in a 200 MeV fission reaction?

14. Show that the fusion of H-2 (2.0141 amu) and H-3 (3.0160 amu) into He-4 (4.0026 amu) and a neutron (1.0087 amu) liberates approximately 18 MeV of energy.

15. Compare (consider advantages, disadvantages, problems, and so forth) the fission and fusion reactions as practical power sources for man.

16. It is found that the dose rate of radiation at a distance of half a meter from a certain x-ray machine is 0.2 rad/hour. If a person working with this machine is not to receive more than 0.02 rad/hour, what minimum distance must he maintain from this x-ray source? (Refer to Graph 10–3.)

17. The National Council on Radiation Protection recommends that an individual's dose of ionizing radiation should not exceed 0.5 rad per year in addition to background and medical exposure. Suppose that your total exposure for a year consists of a two hour experiment in which a sample of Co-60 is used. The dose rate at a normal working distance from the source is found to be 1 rad/hour. (a) From Graph 10–4, what minimum thickness of lead will safely shield this radiation? (b) What is the minimum thickness of concrete for shielding?

SUGGESTIONS FOR FURTHER READING

Andrade, C. da, E. N.: *Sir Isaac Newton: His Life and Work* (Garden City, N.Y., Doubleday Anchor Books, 1958).
———: *Rutherford and the Nature of the Atom* (Garden City, N.Y., Doubleday Anchor Books, 1964).
Asimov, I.: *Asimov's Biographical Encyclopedia of Science and Technology* (Garden City, N.Y., Doubleday and Company, Inc., 1964).
Benade, A. H.: *Horns, Strings, and Harmony* (Garden City, N.Y., Doubleday Anchor Books, 1960).
Bitter, F.: *Magnets, the Education of a Physicist* (Garden City, N.Y., Doubleday Anchor Books, 1959).
Bondi, H.: *Relativity and Common Sense* (Garden City, N.Y., Doubleday Anchor Books, 1964).
Clark, A. C., and the Editors of *Life Magazine: Man and Space* (Life Science Library, New York, Time, Inc., 1964).
Cohen, B. L.: *The Heart of the Atom: The Structure of the Atomic Nucleus* (Garden City, N.Y., Doubleday Anchor Books, 1967).
Gamow, G.: *Gravity* (Garden City, N.Y., Doubleday Anchor Books, 1962).
———: *Thirty Years that Shook Physics: The Story of the Quantum Theory* (Garden City, N.Y., Doubleday Anchor Books, 1966).
Hughes, D. J.: *The Neutron Story* (Garden City, N.Y., Doubleday Anchor Books, 1954).
Jaffe, B.: *Michelson and the Speed of Light* (Garden City, N.Y., Doubleday Anchor Books, 1960).
Kock, W. E.: *Lasers and Holography* (Garden City, N.Y., Doubleday Anchor Books, 1970).
———: *Sound Waves and Light Waves* (Garden City, N.Y., Doubleday Anchor Books, 1965).
MacDonald, D. K. C.: *Faraday, Maxwell, and Kelvin* (Garden City, N.Y., Doubleday Anchor Books, 1964).
Moore, A. D.: *Electrostatics: Exploring, Controlling and Using Static Electricity* (Garden City, N.Y., Doubleday Anchor Books, 1968).
Mueller, C. G., Rudolph, M., and the Editors of *Life Magazine: Light and Vision* (Life Science Library, New York, Time, Inc., 1964).
Romer, A.: *The Restless Atom* (Garden City, N.Y., Doubleday Anchor Books, 1960).
Shamos, M. H.: *Great Experiments in Physics* (New York, Holt, Rinehart, and Winston, 1959).
Stevens, S. S., Warshofsky, F., and the Editors of *Life Magazine: Sound and Hearing* (Life Science Library, New York, Time, Inc., 1965).
Thomson, G.: *J. J. Thomson—Discoverer of the Electron* (Garden City, N.Y., Doubleday Anchor Books, 1966).
Weisskopf, V. F.: *Knowledge and Wonder: The Natural World as Man Knows It* (Garden City, N.Y., Doubleday Anchor Books, 1962).
Wilson, M., and the Editors of *Life Magazine: Energy* (Life Science Library, New York, Time, Inc., 1968).
Wilson, R. R., and Littauer, R.: *Accelerators: Machines of Nuclear Physics* (Garden City, N.Y., Doubleday Anchor Books, 1960).

ANSWERS TO SELECTED PROBLEMS

CHAPTER I

3. yes

5. 20 m/sec

7. (a) 1,000,000
   (c) 1,000,000,000

9. (a) 6.8×10^4
   (c) 18×10^3

11. 0 m/sec

13. ≈ 40 m

15. object reaches distance $= 0$ m

17. $t = 4$ sec

19. -11.4 m/sec

21. $t = 4$ sec

23. (a) 7 km/hr

25. ≈ 6.3 sec

27. 13 m/sec

29. 7 m/sec

33. Decreasing

35. ≈ -30 m/sec

37. $t = 15$ sec

39. 150 m

41. $t = 10$ sec through $t = 15$ sec

43. 16.7 m/sec

45. $t = 17.5$ sec

CHAPTER 2

1. yes

3. 62.5 m

5. 4.5 sec

7. 800 N

11. yes

13. ≈ -39 m/sec

15. 123 m

17. 200 N

19. 1600 J (PE is work done against the force of the spring)

21. $t = 2$ sec and $t = 13$ sec

23. -12.5 m/sec^2

25. $t = 18$ sec

27. 400 m

29. 2×10^4 N

31. (a) 1.96×10^7 Nm
    (c) 2.96×10^7 Nm

33. (a) 6 sec
    (c) 50 m

35. 2 m/sec

37. 4.5 m

39. (a) 40 m/sec and 80 m/sec
    (c) 150 m/sec

CHAPTER 3

1. 4.9×10^3 Nm or J

3. -50 kg m/sec^2 or N

5. 2×10^3 J

7. 2 sec

9. -9.8 m/sec^2

11. 500 N

13. 40 N

15. 2 m/sec^2

17. 2 kg

19. 2.8 N

23. 3.4 m/sec

25. 5.6×10^8 J

27. 6.27×10^{10} J

CHAPTER 4

1. 6.24×10^8 electrons

3. 10,000 V

5. 1.6×10^{-10} J

7. (a) ≈ 8.3 J
   (c) $\approx 5.7 \times 10^3$ V

9. (a) 100 Ω

11. 11 Ω

13. ≈ 21 A

15. (a) 60×10^{-6}
   (c) 20×10^{-6}
   (e) 2.5×10^{-6}

17. 5 Ω

19. (a) No

CHAPTER 5

1. $\dfrac{2\ N}{amp \cdot m}$

5. (a) south

7. 50 Wb/m²

9. clockwise

13. yes

15. 20 webers

17. 4 volts

19. 3.2×10^{-14} N

21. increasing

25. 1.9×10^7 m

27. 8 to 12 seconds

29. 7 sec

31. 2 and 6 seconds

33. 2 and 6 seconds

39. 20 m/sec

41. 20 watts

43. 10 watts

45. 2 amperes

47. (a) 10 volts

49. 20,000 volts

CHAPTER 6

1. (a) 1.17×10^{-5} Hz

3. (a) ≈ 200 m
   (c) 0.1 Hz

5. ≈ 20.4 m/sec
   $L = 0.79$ meters

CHAPTER 7

3. ≈ 40 dB

5. (a) 1/2 m
   (c) 2 octaves

7. 4.8 m

9. 200 Hz

13. 680 m

15. 0.773 m

CHAPTER 8

1. 330 m

3. 2×10^8 m/sec

7. (a) erect, magnified, virtual
   (b) inverted, diminished, real

9. (a) erect, diminished, virtual
   (c) erect, diminished, virtual

15. 7.5 times

17. 2.58×10^8 m/sec

CHAPTER 9

3. 6.63×10^{-26} joules

5. (a) 7×10^{-7} m

7. ≈ 1000 m or 1 km

9. (a) 100 times

11. (a) 5-1; 5-4-3-2-1; 5-3-2-1;
          5-3-1; 5-2-1; 5-4-3-1;
          5-4-1; 5-4-2-1

17.

	IRON-26	COBALT-27	NICKEL-28
first shell	2	2	2
second shell	8	8	8
third shell	14	15	16
fourth shell	2	2	2

CHAPTER 10

3. ≈ 6.4 days

9. (a) $^{232}_{90}\text{Th} \rightarrow {}^{4}_{2}\text{He} + {}^{228}_{88}\text{Ra}$
   (c) $^{14}_{6}\text{C} \rightarrow {}^{0}_{-1}\text{e} + {}^{14}_{7}\text{N}$

11. $^{206}_{82}\text{Pb}$

13. 0.21 amu

17. (a) ≈ 3 cm

INDEX